AF886692

2026

HETI JOURNAL: INTERNATIONAL RESEARCH AND PRACTICE

Print ISSN: 2811-6119
Online ISSN: 2811-6127
ISBN: 978-0-9973368-7-0

The Federation of Horses in Education and Therapy International A.I.S.B.L.

HETI

HEST OG HELSE

HOOFBEATS FOR HEALTH

bridging equestrianism and welfare

Join us from 18th - 22nd May 2027 for HETI's 19th International Congress in Oslo, Norway.

Scan for more information or visit:
www.heticongress.org

The Federation of Horses in Education and Therapy International A.I.S.B.L

Secretariat:
The Federation of Horses in Education and Therapy International A.I.S.B.L.
28 Willow Vale,
Ballybrack,
Co. Dublin
A96C2R4
Ireland

Print ISSN: 2811-6119
Online ISSN: 2811-6127
ISBN: 978-0-9973368-7-0
DOI: https://doi.org/10.58862/YOAR2821

HETI Journal: International Research and Practice

HETI

Table of Contents

HETI

Letter from HETI President

Dear Colleagues and Friends,

It is my pleasure to welcome you to the 2026 edition of the HETI Journal: International Research and Practice. As we continue our mission to support and advance the field of Equine-Assisted Services (EAS), the journal remains an important platform for sharing research, clinical insight and professional knowledge that help strengthen our work.

This edition brings together five articles covering a range of important topics within our field. The issue includes original research on the effects of implementing music therapy concurrently with adaptive riding, caregivers' perceptions of the impact of therapeutic riding on children with autism, and the use of virtual reality to deliver Equine Assisted Services. It also includes a multiple case-study analysis of Autistic experiences with therapeutic horseback riding, as well as a case study on the effects of a short-term intensive hippotherapy programme on a child with cerebral palsy.

Together, these contributions reflect the diversity of approaches currently being explored in Equine Assisted Services, from innovation in service delivery to the lived experience of participants and families, and to focused clinical application. I am especially pleased to see the inclusion of the case study on short-term intensive hippotherapy, as intensive Equine-Assisted Physiotherapy is a topic especially close to my own professional interests.

As always, HETI remains committed to fostering collaboration and knowledge exchange in order to support the continued development of good practice worldwide. I would like to thank the authors, reviewers and editorial team for their valuable work and dedication.

Thank you for your continued commitment to the field.

Warm regards,

Vera Lantelme-Faisan

HETI President

Editorial

Dear Colleagues,

It is with great pleasure that I welcome you to Volume 25, Number 1 (2026) of the HETI Journal: International Research and Practice.

This issue reflects the continuing maturation of research in Equine Assisted Services, with contributions that demonstrate both methodological diversity and a growing commitment to interdisciplinary inquiry. Across this volume, authors from different countries and professional traditions explore how equine-assisted interventions continue to evolve in response to contemporary clinical, educational, and technological challenges.

The international development of the field also continues to highlight an important challenge: terminology. Because services involving equines in education, therapy, rehabilitation, and human development have emerged within different national, professional, and historical contexts, terminology remains diverse across countries and disciplines. In general, the Journal allows a degree of flexibility in terminology, recognizing that many countries continue to use terms that are already established and widely accepted within their own professional frameworks. These national and professional traditions are respected, and for this reason articles may sometimes reflect terminology commonly used within the author's own context.

For clarity in scientific communication, however, HETI encourages authors to use terminology aligned with the HETI Terminology Glossary. These recommendations are informed by the international research initiative: "A Prospective International Study of Terminology - Preliminary Results" and by the ongoing work of the HETI Terminology Taskforce. A more consistent terminology framework strengthens comparability across studies, supports scientific dialogue, and contributes to the continued advancement of the field.

A notable strength of this issue is the breadth of populations and intervention models represented. Several papers focus on neurodevelopmental and neurological conditions, particularly autism and cerebral palsy, while others explore innovation in delivery methods and multimodal intervention design. Together, these contributions illustrate the expanding scientific foundation of Equine Assisted Services and its increasing relevance across health, education, and psychosocial care.

The opening article presents an innovative study from Australia examining the use of virtual reality in equine-assisted learning. This work explores whether virtual horse interaction can reproduce some of the emotional and reflective benefits traditionally associated with direct horse contact. The findings suggest that virtual environments may offer new opportunities for accessibility, especially for individuals who face barriers to participation in conventional programs. This contribution also invites broader discussion about how digital technologies may complement - rather than replace - the relational dimensions of equine-assisted work.

The second article investigates caregivers' perceptions of therapeutic riding for children with autism spectrum disorder. The reported improvements in behavior, communication, confidence, and sensory processing add to a growing body of evidence supporting therapeutic riding as a promising intervention for children with neurodevelopmental differences. Particularly valuable is the caregiver perspective, which helps connect research outcomes with lived family experience.

Another contribution examines the integration of music therapy within adaptive riding lessons, offering an interdisciplinary perspective rarely explored in current literature. This pilot study suggests that combining music and equine movement may enhance attention, self-regulation, and enjoyment, while also raising important questions about sensory integration and co-regulation in equine-assisted settings.

The fourth paper presents a qualitative multiple case study of autistic youths' lived experiences in therapeutic horseback riding. By centering participant voice, the authors move beyond outcome measurement and provide important insight into self-advocacy, accommodation, and the meaning of individualized support within equine-assisted programs. Such contributions are essential as our field increasingly recognizes the importance of participant-centered and neurodiversity-informed research approaches.

Finally, this issue includes a case study from Greece examining a short-term intensive hippotherapy intervention for a child with cerebral palsy. Although limited by sample size, the study offers clinically relevant observations regarding gross motor function, posture, and the potential role of intensive scheduling in therapeutic outcomes. It also reflects the growing international contribution of physiotherapy-based equine-assisted practice to the journal.

Taken together, the papers in this issue point toward several important directions for the future of our field: greater methodological rigor, increased international collaboration, interdisciplinary therapeutic design, and openness to innovation while maintaining careful attention to equine welfare, participant safety, and professional standards.

As Editor-in-Chief, I continue to encourage submissions that combine scientific quality with practical relevance. The strength of the HETI Journal lies in its ability to bring together researchers, clinicians, instructors, therapists, and educators who share a commitment to advancing knowledge in Equine Assisted Services.

I would also like to express my sincere gratitude to our reviewers and editors, whose thoughtful work ensures the academic quality of every issue. Their contribution remains fundamental to the journal's development.

The continued progress of the HETI Journal: International Research and Practice depends on the active engagement of our international community - authors, reviewers, practitioners, and readers alike. I warmly invite you to continue contributing your research, reflections, and scholarly dialogue as we strengthen the evidence base of this important field.

I hope this issue will contribute meaningfully to ongoing scientific and clinical dialogue!

Alexandra N. Stergiou
Editor-in-Chief
HETI Journal: International Research and Practice

Call for Submissions

We are seeking original submissions with the purpose of advancing knowledge in the fields of Equine Assisted Services.

The HETI Journal has several submission categories:

Original Research - An original research study conducted by the author.

Literature Review - A comprehensive or systematic review of research on a specific topic.

Study Protocols - Procedures for proposed original scientific investigations that contribute to the development of Equine-Assisted Services.

Clinical Cases (Case studies) - A presentation of a simple comprehensive case study or multiple case studies.

Theory and Practice - A discussion of a specific topic in the field.

Letter to the Editor - Comments on articles published in the HETI journal or in other journals on topics related to the Equine-Assisted field.

Manuscripts are reviewed through a double-blind peer-review system and submissions are welcome throughout the year.

For more information about article submission visit:

https://hetifederation.org/call-for-journal-submissions/

Contact: Please direct all communication regarding the journal to editors Alexandra Stergiou and Emily Kieson at: editorial@hetifederation.org

Call for Reviewers

The Federation of Horses in Education and Therapy International (HETI) publishes the HETI Journal: International Research and Practice. The focus of this journal is to advance knowledge about equine interactions in human services, such as equine-assisted therapies, equine-assisted activities, and therapeutic riding. Reviewers are needed to ensure the quality and status of the HETI Journal as a peer-reviewed journal. Peer-reviewers conduct blind reviews of manuscript submissions within their area of knowledge as a step in the review process.

Expectations for reviewers are:

• Ability to provide constructive feedback for a scholarly manuscript in terms of its quality and contribution to the field

• Timely and respectful communication with Editorial Team

Requirements for reviewers are:

• Professional experience in an area of equine interactions in human services

• A PHD or Post-Doctoral degree

• Experience in reviewing scholarly texts

• Ability to complete a review within four weeks of receiving a manuscript

If you are interested in contributing your time as a peer-reviewer for the HETI Journal please submit a letter of interest and your curriculum vitae (CV) via email to editorial@hetifederation.org.

Please type "Reviewer Application" in the subject line.

Using Virtual Reality to Deliver Equine Assisted Services

Linda Wong, Associate Professor Claire Henderson-Wilson, PhD, Senior Lecturer Justin Lawson, PhD

School of Health and Social Development, Faculty of Health, Deakin University, Australia

All correspondence and requests for reprints should be addressed to: Linda Wong, at wongli@deakin.edu.au or PO Box 75, Meredith, Victoria, 3333, Australia.

Handling Editor: Emily Kieson
Reviewers: Marilyn Sokolof & Laura Hopson

Volume 25, Number 1, January 2026

DOI: https://doi.org/10.58862/COBI1095

A study using virtual reality to deliver equine-assisted learning in Australia has been undertaken as part of an overall thesis to explore the question "Can interactions with horses in person and in virtual reality improve individuals' health and well-being?". A scoping review, completed for the thesis, identified knowledge gaps, of which two were related to: (1) accessibility and (2) the use of technology. A purposive sample of eight participants participated in the virtual reality study, in which they interacted with horses using virtual reality. A qualitative methodology using Interpretive Phenomenology Analysis was used to understand the participants' experiences with horse interactions in a virtual reality environment.
The findings revealed that virtual reality participants experienced similar benefits as face-to-face participants, identified from sources of evidence in the scoping review: (1) regulating emotion, (2) sense of connection, (3) mindfulness, and (4) physical well-being. The use of virtual reality in equine-assisted learning can be considered to increase the access of equine-assisted learning to a broader range of participants. Ultimately, the outcomes and benefits of delivering equine assisted learning to participants via virtual reality, can improve their overall health and well-being.

Keywords: benefits, horses, virtual reality, well-being

The increased popularity of horses in human service interventions (Amado-Fuentes et al., 2024), referred to as equine-assisted services (EAS) in this article, has several considerations for the triad involved: the participant, horse, and facilitator. The first consideration is protecting the welfare of the participants (Chapman et al., 2024). A preliminary search of the literature identified a broad range of participants, from children to adults, and cohorts with various physical or mental health challenges. A scoping review highlighted that a larger proportion of studies focused on youth with specific challenges. A further complication for the participant was understanding the type of service they believed they were receiving based on service definitions. An example reported by Wood et al. (2021) indicated "three widely used terms of equine therapy, equestrian therapy, and horseback riding therapy, where they were applied inconsistently, defined differently, and used interchangeably" (p. 89). Culture and differences in methodology used across countries also seem to affect terminology (Mattila-Rautiainen et al., 2023). Terminology is so important in safeguarding the participant that if it is not clear and transparent it may confuse, mislead, or fail to protect the participant. Further clarification of international understanding and unison to determine a common denominator of all services is beneficial (Mattila-Rautiainen et al., 2023).

The second consideration from the growth of EAS in the past two decades is the lack of studies focusing on the welfare of horses involved in these services (Baxley et al., 2024). A more recent study highlighted the importance of maintaining the health of horses, selecting suitable horses, and implementing effective care practices to promote positive sentimental positions in EAS (Visser et al., 2025). As with the welfare of participants, terminology and a lack of standard definitions of the working equine present several issues. One issue relates to inconsistencies in academic research, and the second is the influence on the development and implementation of policies intended to improve horse welfare. A lack of policies can lead to misallocation of resources and funding, where certain populations of the working equine can be overlooked (Raw et al., 2024).

The third consideration relates to facilitators' educational requirements. There is a range of certification and accreditation options available to individuals pursuing a career in EAS. Some examples include PATH International based in the US, equine-assisted qualifications based in the UK, and equine-assisted therapy in Australia that provides a diploma of Counselling upon completion. Facilitators/instructors in EAS have indicated a need for research-based education (Rankins et al., 2025). This need is largely due to the growth within the EAS sector, which has increased risk exposure and the requirement for service providers to maintain adequate levels of health, safety, and welfare. There continue to be industry gaps in the lack of training, educational material, and competency (Rankins et al., 2025).

Coupled with growth in the sector in relation to demand and service providers, cost is a financial barrier (Coughlan, 2024; Ballard et al., 2020), which hinders access for many participants. As a service provider in the sector, the main author identified several cost factors that influenced the fee structure to the participants. These are (1) the cost of horse feed, (2) health and maintenance costs, (3) insurance, (4) travel for the participant, (5) property maintenance, and (6) staff. In Australia, unless a participant is covered by the National Disability Insurance Scheme and the service provider is a registered National Disability Insurance Scheme provider, the cost to the participant can be unaffordable.

The delivery of equine-assisted learning (EAL) or therapy typically requires the physical presence of the participant and horse to achieve beneficial outcomes. With the advancement of technology, particularly in the Virtual Reality (VR) arena, it is anticipated that participants who cannot access this type of intervention owing to either physical or mental conditions can experience a similar interaction with horses and still achieve a form of benefit. Some VR studies on horse riding have been reviewed (Ahn et al., 2016; Anderson et al., 2010; Lee and Park, 2014; Satriawan et al., 2022) but nothing has been published, based on the main author's search as of March 2025, with respect to VR simulations to test whether horse interactions in VR can have a similar result to face-to-face interactions. Furthermore, the use of VR in existing EAL approaches may provide alternatives for follow-up sessions, which could increase the accessibility and sustainability of benefits/outcomes for participants.

This article presents the results of a study conducted with participants who interacted with horses in a virtual environment using a VR headset or by watching 360° videos on a computer monitor. The overall aim was to determine whether interactions with horses in VR could improve individuals' health and well-being. The key questions explored were: (1) What was the participant's experience of EAL using VR? (2) What were the changes observed after the EAL sessions with the use of VR? (3) How can VR be incorporated into EAL to increase accessibility and improve outcomes? (4) Is there evidence from this study that warrants further research on the use of VR?

Method

All procedures performed in the studies involving human or animal participants followed the ethical standards of the institutional and/or national research committee and the Helsinki Declaration of 2013. Human ethics approval was granted in August 2019 (DUHREC-2019-255). Animal Ethics approval was granted in February 2020 (AECW-B-B01-2020).

Research Paradigm

Based on the fundamental framework of the main author's beliefs, assumptions, and values, this study adopted an interpretivist paradigm, which is a research philosophy asserting that reality is subjective and socially constructed through individual meanings and interpretations, not an objective, single truth. The authors of this article were motivated to seek individual experiences and consider different interpretations of the social context within EAL to gain a deeper understanding of the meaning and implications for practice (Junjie & Yingxin, 2022). Aligning with this paradigm, the authors took a social constructionist ontological stance, with the view that there is no single objective reality. The belief was that reality is actively created through social interactions, language, and cultural practices. The main author reminded herself to take a critical stance in recognizing her own biases, promoting consideration of the various perspectives of the participants in the studies to understand their world (Burr, 2024). Epistemologically, the authors have taken an interpretivist approach, arguing that knowledge is not passively received from the world but is actively constructed by individuals and groups within specific social contexts. Therefore, in terms of how the findings were interpreted from the studies, the authors posited that knowledge is generated depending on the participant's own sense and experience (Bibi et al., 2022).

Methodology

Three types of qualitative methodologies (Tuffour, 2017) were used for the VR studies: (1) Interpretative Phenomenology Analysis (IPA) – concentrated on examining and interpreting the "lived experiences" of the participants (Alase, 2017), (2) Hermeneutics – adopted two processes to interpret the meaning and significance of the data. One encouraged the participants to interpret the meaning based on their own experiences. As an example, a participant during an interview reflected on the fact that they often asked themselves, "Why am I feeling that way?". Second, the authors interpreted the meaning from any insights the participants offered and how they interpreted the participant's own interpretation (Smith & Osborn, 2003). An example of this process was when the main author reflected on "What did that participant mean when they spoke about how their experiences brought back childhood memories?"; and (3) Idiography – demonstrated the use of this philosophy through VR studies where the experiences and perspectives of each participant were analyzed and described individually. Using divergent and convergent analyses across participants, similarities and differences (Nizza et al., 2021) were identified.

Data analysis method

The exploratory nature of the studies that were participant-oriented (Alase, 2017) and the informed investigation of a smaller, demographically standardized group (Spiers and Riley,

2019) directed the authors' decision to use IPA as the method to analyze the data. The recommended seven-step framework (Smith et al., 2009) was followed: (1) reading and re-reading the transcripts, (2) initial coding, (3) developing emergent themes, (4) searching for connections across emergent themes, (5) moving to the next case, (6) looking for patterns across cases, and (7) taking interpretations to deeper levels.

Rigour of Analysis

Rigor was applied throughout the data analysis process, where data from each participant was analyzed in isolation before it was brought together to be analyzed as a group. The process of generating the transcripts followed three steps: (1) a draft was generated from the video or voice recording, (2) the draft was then checked by the main author to complete the transcription, and (3) the transcripts were sent to the participants for checking. A due date was provided to the participants to return feedback. It was assumed that the participants had no feedback if confirmation was not received by the due date.

Triangulation was used to test the coding of the transcripts using a sample of two transcripts to test the coding. The main author completed the coding first on one transcript, while the other two co-authors completed the coding of the second transcript. The main author documented the coding results from the two transcripts into a spreadsheet to analyze the emerging themes and highlighted any biases that came through. At the conclusion of the analysis, the authors discussed the results in terms of (1) themes at the individual level and (2) themes across the dataset.

Participants

The horses selected for the VR study were experienced in EAL programs and were therefore familiar with the activities involved. The horse breeds that were preselected for filming consisted of four Quarter Horses and one Gypsy Cob in Australia. The horses remained on their home property throughout the filming and at the conclusion of the filming. A description of the filming activities is provided in Appendix A.

A purposive sample of eight participants was recruited so that the significance of the research questions was more applicable to a more closely defined group. The basis for the smaller participant size was due to the selection of IPA, where participants were examined at greater depth over a broader, shallower, and simpler descriptive analysis of many individuals (Pietkiewicz and Smith, 2014).

Theoretical Perspectives

Two theories guided the design of the VR studies: (1) Kolb's (1984) Experiential Learning Theory and (2) Kaplan's (1989) Attention Restoration Theory. Kolb's theory works on two levels: the concept of a learning cycle that encompasses four stages, and the concept of learning styles, of which there are nine (Kolb & Kolb, 2005; Kolb, 2011). The four stages of the learning cycle were used in the design of the VR EAL session plans, where one session followed the four stages of:

1. **A Concrete Experience**: the learner experiences a new situation, or a previous situation is re-encountered to deliver a new experience.
2. **A Reflective Observation** – reflection focusing on any inconsistencies between the experience and the learner's understanding of that experience.
3. **An Abstract Conceptualisation** – based on reflection, the learner has an idea or a change in an existing abstract concept.
4. **An Active Experimentation**: The learner applies learning from the first three steps in their current world to see what results it will bring.

The second level of theory relates to the nine levels of learning styles in which a learner can progress through the four stages of the learning cycle. By combining the four phases of learning and the nine learning styles, an experiential learning framework is formed, as represented in Figure 1.

Figure 1

Kolb's nine levels of learning as it pertains to the learning cycle

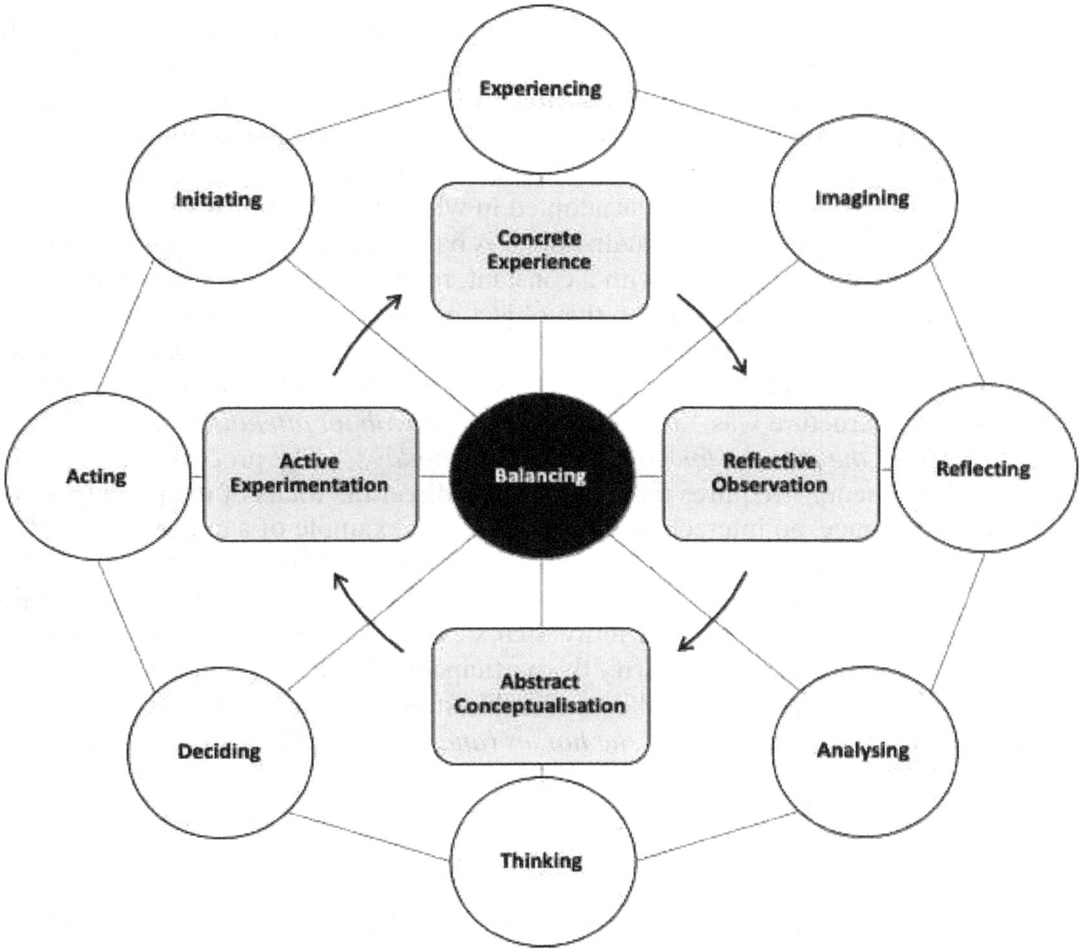

Note. Adapted from "Learning Styles and Learning Spaces: Enhancing Experiential Learning in Higher Education" by D. Kolb & A. Kolb, (2005), *Academy of Management Learning & Education,* Vol. 4, No. 2 (Jun. 2005), pp. 193-212 (http://www.jstor.org/stable/40214287).

Kaplan's Attention Restoration Theory states that people can concentrate better after spending time in a natural outdoor environment, or perhaps even by looking at images depicting nature. Attention Restoration can be beneficial to participants by leveraging a natural outdoor environment to encourage positive thinking (Kaplan 1995; Korpela et al. 2001). EAL is typically delivered in a natural outdoor environment or in an arena, depending on the location and weather conditions. The components of EAL are difficult to differentiate; therefore, a complete experience must be valued. In addition to human-to-human and human-to-horse relationships in EAL, the natural outdoor environment may also be relevant to the overall experience (Hauge et al, 2014). In previous research, participants of EAL stated that being with horses in their natural environment provided them with "attention restoration" elements by assisting them to be relaxed and calm (Burgon, 2014). Although it is more difficult to apply Kaplan's Attention Restoration Theory in the design of VR studies, the natural outdoor environment for all sessions was a key component in the videos.

Measures

The flexible and versatile nature of IPA to understand people's lived experiences (Tuffour, 2017), informed the use of three instruments to collect data. In addition to the traditional method used in IPA, which is an interview guide, artboards and journals were also used. Bartoli (2020) indicated that "there is a growing body of evidence that suggests that visual research methodologies are being applied to qualitative research" (p. 1009). Bartoli (2020) also stated that the combination of visual stimuli alongside interviews can provide a greater depth of understanding of the changes a participant has experienced over time.

A phenomenological interview method was adopted in which the questions were semi-structured in nature. The process of "contextualization" is based on the theory that objects or experiences in the real world tend to relate with a constant, marker, or baseline, which is then linked to a personal story that links the story to that object or experience (Husserl, 1970). Therefore, to examine a participant's particular experience, the context and personal story were considered to determine if the experience had any meaning (Bevan, 2014). An example of a question asked under this structure was *"Tell me how you came about attending your EAL session(s)?"* or *"Describe the process for how you got here today?"*. The process of "apprehending the phenomenon" requires that the question direct the focus of the participant on the experience, in this instance, an interaction with a horse. An example of a question asked under this structure was "*Tell me about (one of) your session(s)?*" or "*Describe what you did at the session?*". The last step in the interview schedule was to "clarify the phenomenon, " and this required identification with an element of the individual's experience and to vary that experience by applying different components to it to clarify the participant's lived experience (Husserl, 1960). An example of a question that was asked under this structure was "*Describe how your experience in these sessions would change if the horses (and virtual reality) were not part of the session?*".

Procedures

Filming of the 360° Videos

Human and Animal ethics approved the filming of horses in Victoria, Australia, interacting with their owners and in their natural environment using a 360° camera. Filming was initiated in September 2020, and the final video was filmed in January 2021. Seven videos were captured using a Garmin 360° camera. The process for transferring the videos to the VR headset consisted of transferring the video file to a computer, editing the file in Adobe Premiere Pro, exporting a VR-formatted file, and uploading that file to YouTube in the correct format for viewing in VR. The end-to-end processing of each video took an average of two–three days. The length of each VR video ranged from 1:30 to 7:30 min, which is within the recommended length for viewing VR videos, as indicated by the Institution of Engineering and Technology (2022). A YouTube channel that was set up for the participants to access the VR videos and a summary of these videos are provided in Appendix B.

Recruitment of Participants

Recruitment was conducted via a third-party not-for-profit Australian organization with approximately 350 staff between 5th October to 16th October 2020, with a start date of data collection from 1st November 2020. Information was provided to a representative from that organization, which included a flyer, plain language statement, and consent form. The representative sent an email to all staff inviting them to participate in the VR studies. Interested participants were asked to return a signed consent form to the main author. To maintain the privacy of the participants, names were withheld from the third-party organization. Upon receipt of signed consent forms from the participants, a welcome pack was posted to eight participants with a pre-paid postage satchel to return their journals at the conclusion of their participation. The pack consisted of: (1) a welcome letter, (2) VR goggles, manufactured by BNext, which were used to hold a mobile phone for viewing the VR videos, (3) instructions on how to use the VR goggles, (4) a journal/notebook, (5) six copies of the artboard, and (6) a guide for how to go through each session. The timetable of the sessions and instructions provided to the participants are outlined in Table 1.

Table 1

VR session timetable and set of instructions to participants

Date	Activity	Contact Point
From 1st November 2020	• Pre-session interview	Zoom
	• Look at the horse artboard provided in your pack (there are several copies for you to use throughout this project) • Feel free to colour in or identify with those images that best resonate with how you are feeling • Write in your journal some thoughts on how you are feeling now	Participant self-assessment
	• Watch the introduction video from the facilitator	YouTube channel
From 15th November 2020	• Watch the session 1A video • Write your observations from the video in your journal • Describe your emotions before and after the video	YouTube channel
From 1st December 2020	• Watch the session 1B video • Write your observations from the video in your journal	YouTube channel

	■ Describe your emotions before and after the video ■ You will have an opportunity to schedule in an online zoom session with the facilitator if you would like to discuss anything	
From 15th December 2020	■ Watch the session 2A video ■ Write your observations from the video in your journal ■ Describe your emotions before and after the video	YouTube channel
From 1st January 2021	Mid-point interview	Zoom
	■ Watch the session 2B video ■ Write your observations from the video in your journal ■ Describe your emotions before and after the video ■ You will have an opportunity to schedule in an online zoom session the facilitator if you would like to discuss anything	YouTube channel
From 15th January 2021	■ Watch the session 3A video ■ Write your observations from the video in your journal ■ Describe your emotions before and after the video	YouTube channel
From 1st February 2021	■ Have a look at the horse artboard again that you referred to in the first week of November ■ Reflect on whether you associated with a different horse(s) now compared to November ■ Write your reflections in your journal ■ Watch the session 3B video ■ Write your observations from the video in your journal ■ Describe your emotions before and after the video ■ You will have an opportunity to schedule in an online zoom session with the facilitator if you would like to discuss anything	YouTube channel

From 15th February 2021	Final interview	Zoom

Participant Sessions and Interviews

Each self-guided session was scheduled two weeks apart and was designed to align with the four phases of learning in Kolb's theory. The viewing of the VR video was aligned to Phase 1, where the participant had a concrete experience via the VR headset. The participants were instructed to write observations in their journal, which aligned with Phase 2, where they were able to reflect on their experience through their observations. The participants were also instructed to describe their emotions before and after the video, which was aligned to Phase 3, where the participant was able to develop their own analysis and conclusions. The interviews allowed the participants to reflect on whether the VR experiences impacted their mindset.

Interviews with each participant were conducted at the following intervals: (1) before the participant watched their first VR video, (2) at the midway point, and (3) after their final VR session. Interviews were conducted and recorded in Zoom, where each interview lasted on average 30 minutes. Recordings ensured that the participants' words were reported and analyzed verbatim. This process ensured that their stories were told through their voices, which were at the core of the IPA (Smith et al., 2021).

Prior to the participants watching the first VR video, they were asked to look at the artboard and color in the images that best represented how they were feeling and write a reflection in their journal. The artboard is a page of horse images set in an outdoor landscape, as per Appendix C. Each image represents a horse activity, and participants could make their own interpretations of the artboard. The final step in the VR studies was for the participants to return their journals in the pre-paid satchel provided in their welcome pack.

Results

The dynamic and flexible nature of the analysis produced Experiential Statements and Personal Experiential Themes (PETs) for each case and Grouped Experiential Themes (GETs) for cross-case analysis (Smith et al., 2021). The results are presented in three parts: (1) Part 1 presents the themes with respect to the participants' experiences with EAL using VR, (2) Part 2 presents the themes that surfaced with respect to the observed changes of the participants after all the VR EAL sessions, and (3) Part 3 presents the experience of using VR headsets in EAL and the need for further studies. Analysis for seven participants was completed due to a participant dropping out during the study.

Results from Part 1 – Experiences with EAL using VR

Experiential statements for each participant were captured using three instruments. PETs were identified for each individual case by collectively analyzing the results of the three instruments. Table 2 summarizes the experiential statements and PETs of each participant.

Table 2

Results from Part 1 – Experiential statements and PETs

	Experience with EAL using VR	
Participant	**Experiential Statements**	**PETs**

	Interviews	Artboard	Journals	Combined instruments
P1	■ Positive recollection of childhood memories ■ Linking a moment in the VR to a personal self-awareness ■ Triggering emotions from muting the sound on the video ■ Ease of use of the VR headset	■ Completed one artboard by placing a sad and happy face on the horses ■ Sad face was placed beside the horse that was tied up and pulling back ■ A happy face was placed beside the horses that were running / playful and the horse standing in pasture watching contentedly	■ Sense of calm ■ Brought about enjoyment and amusement ■ Playfulness of the dogs brought enjoyment ■ Snow and cold brought back memories of living abroad ■ Strong human – horse bond ■ Letting go of concerns to notice what is in front	■ Memory recall ■ Sensory awareness ■ Association with real life
P2		**Dropped out**		
P3	■ Familiarity with surroundings made it comfortable ■ Remembering times with own horse ■ Sense of awareness ■ Curious about what is to come ■ Being in a happy place that provides calm and tranquillity ■ Panning around in a 360 view provided a	■ Completed one artboard by colouring several images ■ Images coloured included: ○ Tree with bird ○ Horse reaching for apple in tree ○ Horse peeking around the tree looking inquisitive	■ Memories of sweet smell ■ Awareness of the removal of a sense ■ Happy place ■ Happy and joyful ■ Calm, relaxed and grounded ■ Wellness and me time ■ Peaceful ■ A way of expression ■ Without sound - video didn't make sense whereas	■ Positive influence on emotions ■ Reflective of life ■ In the moment

	■ Such enjoyment in watching an animal (dog) being by itself ■ Well managed horses ■ Connected to the dog videos more than the horses ■ Emotionally engaged ■ Disconnected from the aerial video ■ Delighted and uplifting	them in the journal ■ Identified three negative images ■ Other images of happy and healthy ponies were quickly coloured based on concern of own ponies being susceptible to foundering	■ Dog videos were high energy and engaging ■ Horses felt aloof ■ The days between Christmas and New Years is anchorless ■ Laughed at own obsessions	
P6	■ Enhanced the sense of imagination ■ Brought back memories of owning a horse and living abroad ■ Sensory experience ■ Created space to reflect on life and relationships with people ■ Multiple viewings added opportunities to observe new things ■ Quality of videos were good ■ Worried about following the process at the start but eased into the flow to enjoy the process	■ Completed an artboard for each video session – six in total ■ Images selected over time changed from the more stressed and lonely horses at the start to the one picking an apple off the tree for the last two sessions	■ Concerned about following the process ■ Observation that the horses trust and respect human ■ Reflected on how much more was heard in the sounds when just listening versus bringing the visual. ■ Reflection on sound – or lack of ■ Pleasant repetitive sound of the grooming ■ Soothing sounds ■ Awaken the desire to watch other horse videos on Youtube for enjoyment	■ Self-conscious ■ Connected ■ Reflective learning

	■ More engaged in the videos with the horses ■ Videos without horses were nice to watch and listen to, but not as engaging			
P7	■ Structure provided focus ■ Scenes creates a sense of peaceful ■ Found humour in watching the horses interact ■ Connection to nature ■ Relate to the temperatures based on what was in the video ■ Enjoyment in working out the horse characters ■ Can imagine being there ■ Brought back fond memories of home and owning horses as a child	■ Cutout the horse images and pasted in journal twice ■ At start of sessions, cutout and coloured horses grazing on grass and drinking from stream ■ At mid-way point, picked horses, cutout and coloured two horses. One picking apple off tree and the other pawing at water playfully.	■ Feeling relaxed and content ■ Peaceful and calm settings ■ Strong connection to being aware of the five senses ■ The snow and cold brought about feeling of isolation ■ Activity of feeding out hay create a sense of feeling itchy ■ Countryside sounds and nature ■ Being pampered	■ Connection to nature ■ End-to-end experience ■ Creativity and imagination
P8	■ Enjoyed the structure and felt organised ■ Checklist of prompts ■ Experimented with the immersive nature of the VR	■ Completed one artboard by circling the images and writing words ■ Pre starting the sessions, circled the horse in the shelter. Wrote	■ Laughed out loud ■ Mind map to being well ■ Watch horses with satisfaction ■ Observe the different personalities	■ Being an observer ■ Engaged ■ Meaning of life

■ Relaxing and enjoyable to watch ■ The camera where it was up close felt more engaging ■ Being an observer didn't feel as engaged ■ Good amount of time with the videos ■ Interaction between the horses and with the facilitator	that being on own is okay. ■ Post the sessions, circled the horse picking an apple from the tree. Wrote positive thoughts.	■ Reflection of own happiness with feeding own animals ■ Video 3A – felt like more of an observer ■ Video 4A – Drone was cool and good to get the lay of the property. ■ Video 4B – Calm and peaceful activity. Nice to get up close with the horses again.

A cross-case analysis of participants' experiences with EAL through VR determined four GETs: 1) sensory processing, 2) reflective learning, 3) influencing emotions, and 4) degrees of connection. Descriptions of each GET along with a participant quote supporting each GET are provided in the following paragraphs:

GET 1.1: Sensory processing

All participants had heightened awareness of at least one of their senses, primarily their sense of hearing and sight. Observations were made about other senses in that participants could imagine smells or what certain things felt like to touch. An example is provided by participant 1, recalling a sense of smell:

> It's quite interesting the effects VR has [sic] on the actual mind itself. And (…) there's times [sic] where my mind's tricking me to think that I can actually smell the hay and I can almost smell the horse and the fresh air and stuff like that.

GET 1.2: Reflective learning

To assist in navigating the VR EAL sessions, the structure and instructions provided to the participants introduced learning opportunities both during and after the viewing. Reflections from the participants indicated a realization of or connection to a memory from their past. An example is provided by participant 5.

> I do remember (…) that I responded in my interpretation of images was more about the animal. Like, oh, my goodness, it's out in a pasture of grass. It'll be [eating too much green grass and get sick] soon (….) I think I was interpreting everything from the perspective of where I was at the time.

GET 1.3: Influencing emotions

The VR videos were designed with different scenes, which had varying degrees of influence on how the participants felt while watching the video. Participants identified a sense of enjoyment or happiness. An example is provided by participant 7.

> And I love horses with those big eyes, . . . in the real close-up [video] I swear I can see the horse's eyes . . . like the one that was chewing on the vehicle right in front of the camera. That

made me laugh so much. It was so funny. It's brought quite a lot of emotions into it and it's very peaceful. But there's also that element of humour as well.

GET 1.4: Degrees of connection

The level of engagement or connection that the participants had with each VR EAL session varied depending on what was viewed and heard in the video. The proximity of the camera, sound, and presence of horses seemed to have a key influence on participant engagement. An example was provided by Participant 1.

> But other than that, the headset wasn't too heavy or inconvenient and at times you could be immersed in it so much that you're kind of like going to a different world.

Results from Part 2 – Observed changes of the participants post the VR EAL sessions

The analysis conducted for each participant captured experiential statements across the three instruments. After capturing the experiential statements, the PETs for each individual case were identified by collectively analyzing the results from the three instruments. Table 3 summarizes the experiential statements and PETs of each participant.

Table 3

Results from Part 2 – Experiential statements and PETs

Participant observations post their EAL sessions via VR				
Participant	**Experiential Statements**			**PETs**
	Interviews	**Artboard**	**Journals**	**Combined Instruments**
P1	■ Acceptance of circumstances ■ Positive influence on physical and mental well-being	■ Went from feeling of being tired to one of playfulness and contentment	■ Learned more about inquisitive nature of horses ■ Ability to tune out a negative thought and be in the moment	■ Influence on well-being ■ Mood enhancer ■ In the moment
P2		Dropped out		
P3	■ Bank of memories ■ Being in nature and with horses is something to get back ■ Take the blinkers off to take in more	■ Went from sitting on the sidelines to being an active player	■ Realisation for oneself that being in nature and the reliance on all the senses is needed to be present and in the moment	■ Personal well-being ■ Contributor ■ Connection to nature

	■ Realisation of how each sense can have an impact of the whole experience ■ Make time for things you can benefit from			
P4	■ Apprehension at start but relaxed into it ■ Focus was required with busy lifestyle ■ Learned that horses are curious	■ Moved from the more stressed, anxious and lonely one to calmer images	■ Sound has a profound impact on the experience ■ Connection to sun, birds chirping and sense of fresh air ■ Time to relax without interruptions or rushing	■ Focus ■ Taking time out ■ Regulate stress
P5	■ Provided time to reflect on life ■ A valuable experience that did something	■ Amused with own obsession of the horse images in the artboard with own reality	■ Observations means being in the moment ■ Calm energy is a healing energy ■ Writing allowed for realisation of being distressed and distracted. The admittance in writing then allowed for calm and clearness ■ Uplifting when an animal chooses to be with us. ■ Conscious effort to focus and rethink	■ Reflection ■ Link to reality ■ Healing

P6	■ Motivated and enthusiastic compared to four months ago ■ Easing of agitation with people	■ Before associated with being tied up and stress ■ After associated with calm and resting	■ Worried less about what people think ■ It is okay to not be liked or like everyone	■ Influencing emotions ■ Acceptance of oneself ■ Positive mood
P7	■ Horses are like people with different personalities	■ More playful and inquisitive towards the end	■ Strong connection to the outdoors, blue skies and the want for country air	■ Learning ■ Connection to Country ■ Curious
P8	■ A great mechanism for lifting the mood ■ Felt better mentally ■ Want to be outside and in the bush	■ Moved from feeling alone and a bit flat to being inquisitive and positive	■ Tendency to move from a negative or neutral state to a happy state	■ Improve mood ■ Desire for outdoors ■ Varying engagement levels

Conducting a cross-case analysis of the observed changes after the participants' experiences with EAL through VR, three GETs were determined: 1) impact on well-being, 2) awareness of nature, and 3) awareness of being present. Descriptions of each GET along with a participant quote supporting each GET are provided in the following paragraphs:

GET 2.1: Impact on well-being

The observed changes in participants after the VR EAL sessions identified a greater awareness of their mental and physical well-being. However, awareness went deeper to influence mindset shifts and improve well-being. An example was provided by participant 8.

> I think I feel comfortable around horses. I think that's kind of reinforced that and how nice it is being outside and being in the bush. That was a nice experience as well.

GET 2.2: Awareness of nature

There was one consistent element across all the 360° videos, which was they were filmed outdoors in nature. Participants often commented on their connection to nature through visual or acoustic elements, which had an impact on their feelings or thoughts. A quote from participant 1 is provided.

> After I watch the video, I do get a strong urge to go for those walks and make sure that I'm walking in the green and next to the river.

GET 2.3: Being in the moment

Participant reflections after their VR EAL sessions indicated greater clarity of their state of awareness and their need for focus. Awareness of senses, emotions, and well-being, identified in

Part 1, demonstrated how participants applied their awareness to be more mindful. This awareness may have prompted the participants to introduce meditation to their lifestyles. An example was provided by participant 4.

> It was very hard for me just to sit and . . . watch something because I'm a very busy person and I'm always doing something . . . I don't have a long attention span. So, it was very challenging for me . . . to focus on these videos and put all my thoughts into just watching them and seeing what would happen.

Results from Part 3 – The use of VR in EAL and the need for further studies

Analysis of the participants' experiences using VR technology identified six factors that had the greatest influence on the participants' experience: 1) age profile, 2) IT literacy, 3) elements in the video, 4) sound in the video, 5) location of the camera, and 6) length of the video. These factors are described in more detail in Table 4 and can be extended to the broader range of EAS that consider the use of VR.

Table 4

Factors influencing the VR experience

Factor	Implications/Considerations
Age profile of participant	The use of virtual reality in EAL can be an alternative option for those aged under 50. The use of VR for those between the age of 51 to 70 requires consideration of the impact of the other factors.
Technical literacy	Those who work in Information Technology or have a medium to high competency with technology are more likely to adopt the use of a virtual reality headset. Considerations for more advanced and expensive virtual reality headsets should be made for participants who find technology challenging or need additional support.
Elements in the video	The consistent elements across all the videos is that they were filmed in a natural outdoor environment where the animals (horses and dogs) were free to choose to interact or not. It seems that at a minimum, VR EAL sessions should be filmed to be as close an experience to that of being face-to-face to evoke stronger emotions and connection for the participant.
Sound	The removal of sound from a video had a much greater impact than the removal of the video image. A common observation, made by the participants, was their connection to nature via the sounds of birds chirping, wind swirling, snow falling, leaves rustling, horses blowing and other noticeable sounds of the natural outdoors.
Location of camera	The closer the camera was to the elements being filmed, in particular the animals, the higher the engagement level.

	Therefore, careful consideration should be taken with not only the elements being filmed, but the angle and placement of the camera.
Length of video	If using strictly VR headsets to provide an immersive experience, it is advisable to keep the videos to recommended lengths of no more than 15 minutes for top-end headsets and no more than five minutes for the less advanced technology.

Discussion

Strengths of the VR case studies

The findings revealed that VR participants experienced similar benefits as face-to-face participants identified from sources of evidence in the lead author's scoping review: (1) regulating emotion, (2) sense of connection, (3) mindfulness, and (4) physical well-being. Figure 2 demonstrates the similarities between the themes of the VR studies and the scoping review.

Figure 2

Similarity of benefits between VR participants and Scoping Review findings

			Themes from Scoping Review relating to key benefits			
			Regulating emotion	Sense of connection	Mindfulness	Physical wellbeing
Themes from VR Studies	Prior to completing all EAL session	Sensory processing		X	X	
		Reflective learning			X	
		Influence emotions	X			
		Degrees of connection		X		
	Post completion of all EAL sessions	Impact on wellbeing				X
		Awareness of nature		X		
		Being present			X	

The VR studies addressed two limitations identified from the scoping review: (1) transparency of the timeframe and number of sessions in the sources of evidence and (2) a broader range of participants.

Timeframe and number of sessions

Transparency was provided in the timeframe of delivery, along with the number of sessions completed by each participant. Participants were influenced in as little as one session, but the real impact seemed to occur after four sessions.

Broader range of participants

The VR studies addressed this limitation by using a third party organisation, where the researchers advertised and recruited staff to participate. The findings indicate that consistent design and delivery methods could be applied to all participants. Future studies should consider qualitative methods for standard adults and quantitative methods for youth or cohorts with specific needs.

Limitations from the VR case studies

A key limitation was that at the time of delivering the VR studies, Victoria, Australia was amid Covid-19 restrictions for many weeks in lockdown. The participants completed their sessions during the lockdown, which may have influenced their VR experience. Participant interviews indicated that being involved in VR studies provided an escape from the stresses of lockdown.

Another limitation is the sample size. Based on the scoping review, qualitative studies ranged from six (Walsh & Blakeney, 2013) to 66 (Adams et al., 2015) participants. The cohort studies ranged from 17 (Battisti et al., 2017) to 64 (Murphy et al., 2017) participants. The three RCT studies used 26, 75, and 133 (Frederick et al., 2015; Hauge et al., 2014; Ho et al., 2017) participants. One case-control study included 39 participants (Boshoff et al., 2015). While the VR case studies did not recruit many participants, the choice to use IPA did not require the use of large sample sizes. Smith and Osborn (2003) recommended that research students using IPA for the first time should consider three as the sample size.

Conclusion

Based on the benefits that the participants experienced from the VR studies, further investigation is warranted. The VR studies in this article highlighted that VR can assist in the awareness and recognition of a participant's current state of emotions. Coupled with structured facilitation and instructions on reflections, awareness can be heightened and monitored by participants. Videos at 360° have a sense of engagement through the activation of the senses, which in turn connects participants to nature, bringing about engagement through what they saw and heard. Greater learning opportunities should be introduced through 360° videos and VR goggles by creating experiences that require the focused use of hearing and sight. The addition of appropriate facilitation and instructions for each VR EAL session could be an effective learning mode for many participants. Creating the presence of physical activities and interaction via VR EAL sessions is more difficult; however, using active videos in VR can provide participants with a desire to be more physically active or create an awareness of their lack of well-being.

VR EAL can provide benefits to participants and should be considered an alternative for those who cannot physically attend in person or for those who are very anxious about attending in person. There is greater potential in the learning and mindfulness space; therefore, further exploration of in-class instruction and application for students with learning challenges is recommended.

Data Availability Statement: Data available on request from the corresponding author.

Conflict of Interest: The authors declare no conflict of interest.

Funds: No funds received.

Author Contributions: Conceptualization, LW; methodology, LW; formal analysis, LW; investigation, LW; resources, LW; data curation, LW, CHW, JL; writing-original draft preparation, LW; writing-review and editing, LW, CHW, JL; project administration LW. All authors have read and agreed to the published version of the manuscript.

Statement: The study was conducted according to the guidelines of the Declaration of Helsinki and approved by the Institutional Review Board of Deakin University (2019-255) and by Animal Ethics Committee Wildlife-Burwood (AECW-B): B01-2020.

References

Adams, C., Arratoon, C., Boucher, J., Cartier, G., Chalmers, D., Dell, C. A., Dell, D., Dryka, D., Duncan, R., Dunn, K., Hopkins, C., Longclaws, L., MacKinnon, T., Sauve, E., Spence, S., & Wuttunee, M. (2015). The helping horse: How Equine Assisted Learning contributes to the wellbeing of First Nations youth in treatment for volatile substance misuse. *Human-animal interaction bulletin*, *1*(1), 52–75.

Ahn, S. J., Bostick, J., Ogle, E., Nowak, K. L., McGillicuddy, K. T., & Bailenson, J. N. (2016). Experiencing nature: Embodying animals in immersive virtual environments increases inclusion of nature in self and involvement with nature. *Journal of Computer-Mediated Communication*, *21*(6), 399-419. https://doi.org/10.1111/jcc4.12173

Alase, A. (2017). The Interpretative Phenomenological Analysis (IPA): A guide to a good qualitative research approach. *International Journal of Education and Literacy Studies*. https://doi.org/10.7575/aiac.ijels.v.5n.2p.9

Anderson, F., Annett, M., Bischof, W. F., & Boulanger, P. (2010). Virtual Equine Assisted Therapy. [Paper presentation]. IEEE Virtual Reality Conference, VR 2010, Waltham, Massachusetts, USA. https://doi.org/10.1109/VR.2010.5444776

Amado-Fuentes, M., Denche-Zamorano, A., Barrios-Fernandez, S., & Gozalo, M. (2024). Bibliometric analysis on Equine-Assisted Interventions. *Animals (2076-2615)*, *14*(12), 1776. https://doi.org/10.3390/ani14121776

Ballard, I., Vincent, A., & Collins, C. (2020). Equine Facilitated Psychotherapy with young people: Why insurance coverage matters. *Child & Adolescent Social Work Journal, 37*(6), 657-663. https://doi.org/10.1007/s10560-020-00712-1

Bartoli, A. (2020). Every picture tells a story: Combining interpretative phenomenological analysis with visual research. *Qualitative Social Work, 19*(5-6), 1007-1021. https://doi.org/10.1177/1473325019858664

Battisti, H. E., Battisti, F. L., & McAbee, R. (2017). Does long-term Equine Assisted Learning have an effect on childhood weight management? *Journal of Obesity & Weight Loss Therapy, 7*(3), 338. doi:10.4172/2165-7904.1000338

Baxley, B. H., Rodriguez-Mori, H., & Anderson, N. C. (2024). Welfare and stress of horses used for Equine-Assisted Services: A systematic review. *Applied Animal Behaviour Science*, *280*, n.p. https://doi.org/10.1016/j.applanim.2024.106424

Bevan, M. (2014). A Method of Phenomenological Interviewing. *Qualitative Health Research, 24*(1) 136–144. https://doi.org/10.1177/1049732313519710

Bibi, H., Khan, S., & Shabir, M. (2022). A Critique of research paradigms and their implications for qualitative, quantitative and mixed research methods. *Webology, 19*(2), 7321-7335.

Boshoff, C., Grobler, H., & Nienaber, A. (2015). The evaluation of an equine-assisted therapy programme with a group of boys in a youth care facility. *Journal of Psychology in Africa, 25*(1), 86-90.

Burgon, H. L. (2014). Horses, mindfulness and the natural environment: Observations from a qualitative study with at-risk young people participating in therapeutic horsemanship. *International Journal of Psychosocial Rehabilitation, 17*(2), 51-67.

Burr, V. (2024*). Social constructionism* (4th ed.). Routledge. https://doi.org/10.4324/9781003335016

Chapman, M., Fenner, K., Thomas, M. J. W., & Thompson, K. (2024). Stakeholder views on the potential benefits and feasibility of an equestrian industry-specific health, safety and welfare management system. *Animals (2076-2615), 14*(23), 3450. https://doi.org/10.3390/ani14233450

Coughlan, J. (2024). Exploring the effectiveness of an RDA riding centre in understanding and meeting the needs of autistic young people and adults. *Good Autism Practice*, *25*(1), 35–41.

Frederick, K. E., Ivey Hatz, J., & Lanning, B. (2015). Not just horsing around: The impact of Equine-Assisted Learning on levels of hope and depression in at-risk adolescents [Report Author abstract]. *Community Mental Health Journal, 51*(7), 809-817. https://doi.org/10.1007/s10597-015-9836-x

Hauge, H., Kvalem, I. L., Berget, B., Enders-Slegers, M.-J., & Braastad, B. O. (2014). Equine-assisted activities and the impact on perceived social support, self-esteem and self-efficacy among adolescent – an intervention study. *International Journal of Adolescence and Youth, 19*(1), 1-21. https://doi.org/10.1080/02673843.2013.779587

Ho, N. F., Fung, D. S. S., Zhou, J., & Kua, P. H. J. (2017). Equine-assisted learning in youths at-risk for school or social failure [Article]. *Cogent Education, 4*(1). https://doi.org/10.1080/2331186X.2017.1334430

Husserl, E. (1960). *Cartesian meditations*. M. Nijhoff.

Husserl, E. (1970). *The crisis of European sciences and transcendental phenomenology*. Northwestern University Press.

Institute of Engineering and Technology. (2022). *Safeguarding the metaverse*. https://www.theiet.org/media/9836/safeguarding-the-metaverse.pdf

Junjie, M., & Yingxin, M. (2022). The discussions of positivism and interpretivism. *Online Submission, Global Academic Journal of Humanities and Social Sciences, 4*(1), 10-14. https://eric.ed.gov/?id=ED619359.

Kaplan, R., & Kaplan, S. (1989). *The experience of nature: A psychological perspective*. Press Syndicate of the University of Cambridge.

Kaplan, S. (1995). The restorative benefits of nature: Toward an integrative framework. *Journal of Environmental Psychology, 15*(3), 169-182. https://doi.org/10.1016/0272-4944(95)90001-2

Kolb, D. (1984). *Experiential learning: Experience as the source of learning and development*. Prentice Hall.

Kolb, A. Y., & Kolb, D. A. (2005). Learning styles and learning spaces: Enhancing experiential learning in higher education. *Academy of Management Learning & Education, 4*(2), 193-212.

Kolb, D. A. (2011). Kolb Learning Style Inventory. Retrieved from https://learningfromexperience.com/themes/kolb-learning-style-inventory-lsi/

Korpela, K. M., Hartig, T., Kaiser, F. G., & Fuhrer, U. (2001). Restorative experience and self-regulation in favorite places. *Environment and Behavior, 33(4),* 572–589. https://doi.org/10.1177/00139160121973133

Lee, D., Lee, S., & Park, J. (2014). Effects of indoor horseback riding and virtual reality exercises on th dynamic balance ability of normal healthy adults. *Journal of Physical Therapy Science, 26*(12), 1903 1905. https://doi.org/10.1589/jpts.26.1903

Mattila-Rautiainen, S., Brennan, R., Emond, N., Horne, V., Volpe, G., Arrieta, K., Stergiou, A (2023). A prospective Internation Study of Terminology in human-equine interactions. *HETI Journal*, *22*(1):28-42. https://doi.org/10.58862/NJBV1737

Murphy, L., Wilson, J., & Greenberg, S. (2017). Equine-Assisted Experiential Learning in occupational therapy education . *Journal of Experiential Education, 40*(4), 366-376. https://doi.org/10.1177/1053825917712732

Nizza, Isabella and Farr, Joanna and Smith, Jonathan A. (2021) Achieving excellence in interpretative phenomenological analysis (IPA): Four markers of high quality. *Qualitative Research in Psychology, 18*(3), 369-386. https://doi.org/10.1080/14780887.2020.1854404

Pietkiewicz, I., & Smith, J. (2014). A practical guide to using Interpretative Phenomenological Analysis in qualitative research psychology. *Psychological Journal*, 20, 7-14. https://doi.org/10.14691/CPPJ.20.1.7

Rankins, E. M., Reega, S. J., Rudd, C., & Peters, B. C. (2025). Horse behavior and training in adaptive/therapeutic riding: A survey of instructors' perceptions and knowledge in the United States. *Applied Animal Behaviour Science, 282,* 164290 https://doi.org/10.1016/j.applanim.2024.106490

Raw, Z., Collins, J. A., & Burden, F. A. (2024). What is a working equid? Analysis of current terminology and a suggested definition. *Animals (2076-2615), 14*(14), 2026. https://doi.org/10.3390/ani14142026

Satriawan, A., Trusaji, W., Hasanuddin, M. O., Rahadini, S. S., Selekta, M. C., & Sungkar, E. (2022). Design of virtual reality-based hippotherapy simulator exergaming software and its controller for rehabilitation of children with cerebral palsy in Indonesia: An engineering concept. *Designs*, *6*(5), 76. https://doi.org/10.3390/designs6050076

Smith, J. A., Flowers, P., & Larkin, M. (2009). *Interpretative phenomenological analysis: Theory, method and research.* London: Sage Publications Ltd.

Smith, J.A., Flowers, P, and Larkin, M. (2021). Collecting data. In A. Maher & E. Carter (Eds.), *Interpretative Phenomenological Analysis: Theory, Method and Research* (2nd Edition, pp. 53-74). Sage publications.

Smith, J. A., & Osborn, M. (2003). Interpretative phenomenological analysis. In J. A. Smith (Ed.), *Qualitative psychology: A practical guide to research methods* (pp. 51–80). Sage Publications, Inc.

Spiers, J., & Riley, R. (2019). Analysing one dataset with two qualitative methods: The distress of general practitioners, a thematic and interpretative phenomenological analysis. *Qualitative Research in Psychology, 16*(2), 276-290. https://doi.org/10.1080/14780887.2018.1543099

Tuffour I (2017). A critical overview of interpretative phenomenological analysis: A contemporary qualitative research approach. *Journal of Health Communication, 2*(4), 52. https://doi.org/10.4172/2472-1654.100093

Visser, E. K., Jens, A. L., Nieuwe Weme, L. E., Spaapen, A. A., Maarleveld, K. N., Enzerink, K. H., Tromp, P. N., & Haven-Pross, S. C. (2025). Assessing equine behavioural responses in Equine-Assisted Services: A field study analysis. *Animals (2076-2615), 15*(5), 671. https://doi.org/10.3390/ani15050671

Walsh, K. M., & Blakeney, B. A. (2013, 06 / 01 /). Nurse presence enhanced through Equus. *Journal of Holistic Nursing, 31*(2), 121-128. https://doi.org/10.1177/0898010112474721

Wood, W., Alm, K., Benjamin, J., Thomas, L., Anderson, D., Pohl, L., & Kane, M. (2021). Optimal terminology for services in the United States that incorporate horses to benefit people: A consensus cocument. *Journal of Alternative & Complementary Medicine, 27*(1), 88-95. https://doi.org/10.1089/acm.2020.0415

Appendix A – Sample of VR filming activities

Treatment Group	Activity	Description	# Horses	# Participants	Duration	Frequency
VR case study participants	Observing horses in their environment – minimum of 5 metres away	■ Five horses in their natural environment going about their day as usual ■ Camera attached to fence in paddock to film without any human interaction	5	0	60 minutes	Once
	Brushing and patting a horse	■ Camera mounted on a tripod in the paddock ■ Owner to brush a horse from head to toe ■ Pick feet ■ Comb mane and tail	1	1	30 minutes	Once
	Halter a horse	■ Camera mounted on a tripod in the paddock ■ Film owner demonstrating how to approach a horse and then putting a rope halter on the horse's head ■ Demonstrate how to remove the halter	1	1	15 minutes	Once

	Feeding the horses	<ul><li>Filmed in three parts</li><li>Part 1 in the shed without any horses</li><li>Film owner preparing the feeds</li><li>Part 2 is bringing the hard feed to the horses in buckets</li><li>Film the owner feeding the horses in their paddock</li><li>Entering the gate, putting the buckets on the ground and leaving the paddock</li><li>Part 3 is preparing the haynets and putting them in the paddock</li><li>Film the owner filling the haynets, putting them in the wheelbarrow, going to the paddocks and placing the haynets in the paddocks</li><li>Film the horses behaviour as they eat</li></ul>	5	1	30 minutes	Once

Appendix B – Videos provided to VR case study participants

Video with YouTube link	**Brief Description**
Intro	The participants were asked to capture their thoughts in their journal prior to watching the scheduled videos. There was a welcome video that was provided.
1A	In this video, a dog runs in view and the facilitator provides instructions to the participants at the start in terms of what they need to think about and capture in their journals. Several scenes have been joined together to introduce the horses in an outdoor environment where they are left to behave as they normally would. There is no talking in these scenes, but there is human interaction that can be seen with the horses.
1B	In this video, a narration is provided in the voice of one of the horses. It is feeding time and a narration is provided that explains the feeding process.
2A	In this video, the participant is introduced to the John Deere Gator utility vehicle, which is like a golf cart. The horses can be seen from the viewpoint of someone sitting in the gator. The facilitator is seen brushing one of the horses.
2B	In this video, there is snow! The participant can see a black dog running in the snow and a person is seen heading to the paddocks to check on the horses.
3A	In this video, dogs are seen playing in a large body of water on a sunny day. One dog is seen to jump into the water to fetch sticks, but the other dog does not enter the water.
3B	In this video, the facilitator is seen in the paddock interacting with the horses. He puts a halter on the black and white horse. The participant can see the facilitator brushing this horse and applying fly spray.
4A	In this video, there is no sound and footage was taken with the 360° camera placed in a drone. As such, the video is a little shaky in parts. The participant can watch the reactions of the horses.
4B	In this video, the 360° footage is captured by wearing the camera on the facilitator's head. The participant can be virtually close to the activities of picking up poo and brushing the horses.

Appendix C – Artboard

Riding Towards Improved Outcomes: Caregivers Perceptions of the Impact of Therapeutic Riding on Children with Autism. A pilot study

Abigail Swilley, OTD, Elizabeth K. Schmidt, PhD, OTR/L

College of Mathematics, Sciences & Health Professions, Department of Occupational Therapy, Lincoln Memorial University

All correspondence and requests for reprints should be addressed to: Abigail Swilley, at abigail.swilley@lmunet.edu or Lincoln Memorial University, 9731 Cogdill Road Knoxville, TN 37932

Handling Editor: Emily Kieson
Reviewers: Emili B. Alexander & Jo Anne Miller

DOI: https://doi.org/10.58862/IGTF2906

Children with Autism Spectrum Disorder (ASD) may experience social challenges and behaviors that can have a significant impact on their quality of life and how they function within their environment. Equine-assisted therapy has been shown to have positive effects on children with ASD and can foster a sense of connection and encourage self-expression through interaction with their equine partner. This study assessed caregivers' perceived effectiveness of Therapeutic Riding (TR), a type of equine-assisted therapy, for children with ASD. Caregivers were asked to reflect on behavioral, social, sensory, confidence, and activities of daily living changes that have occurred since their children first began TR, using a quasi-experimental survey design, from January to February 2025. We hypothesized that caregivers would perceive positive outcomes in all domains. Seventeen caregivers responded to the survey, with the most notable improvements reported in the areas of behavior (88%), self-control (82%), sensory processing (76.5%), social communication (82.4%), independence in daily activities (64.7%), confidence at home (76.4%), and when trying new things (82.4%). However, this study was significantly limited by the nature of quasi-experimental designs (i.e., no control group for comparison) and the lack of a standardized measurement tool. These findings suggest that TR may be used as a promising intervention to improve outcomes in children with ASD, but future research is needed.

Keywords: Autism, behavior, communication, confidence, improvement

Therapeutic riding (TR) is a type of equine-assisted therapy (EAT) that allows individuals with disabilities to engage in horseback riding to achieve a wide variety of therapeutic outcomes primarily related to flexibility, balance, muscle strength, coordination, self-esteem, and confidence (PATH International, 2022). Specifically, TR utilizes therapeutic activity while on horseback to foster communication and interactions between riders and their equine partners

(Zhao et al., 2021). The riders' natural interest in horses is thought to motivate their participation in the TR. According to Self-Determination Theory, intrinsic motivation is strengthened when activities support autonomy (i.e., opportunities for choice), competence (i.e., developing and demonstrating abilities), and relatedness (i.e., feeling connected to others) (Deci & Ryan, 1985). TR commonly supports these needs by allowing riders to make choices during activities (autonomy); providing structured, progressively challenging riding tasks that enable skill development (competence); and facilitating supportive relationships with the horse, instructor, and volunteers (relatedness). As riders experience success and build competence, these accomplishments can serve as mastery experiences, which Bandura identified as the strongest source of self-efficacy, strengthening individuals' belief in their own abilities and supporting confidence that may generalize to other activities outside of TR (Bandura, 1977). Thus, it is unsurprising that TR can have a variety of positive effects on the physical, intellectual, psychosocial, speech, attention, and sensory integration domains (Zhao et al., 2021).

A non-profit organization located in East Tennessee takes pride in its establishment providing therapeutic experiences to individuals with disabilities or other challenges through the healing power of a horse. These therapeutic experiences include engagement in equine activities, both on and off horseback, to achieve goals and promote health and well-being among participating riders. At this facility, riding sessions may occur in groups or private sessions depending on the level of support for individual needs. Group sessions had a maximum of four riders, and private sessions included only one rider. Private sessions are a good option if the rider needs an increased amount of one-on-one attention from the instructor (i.e., increased challenging behaviors that need to be directly managed by an instructor for safety). Riding sessions typically involve a Certified Therapeutic Riding Instructor (CTRI), a horse leader, side walker(s), and rider. To become a CTRI, instructors must complete the following three phases of training. First, applicants must complete the standard course and required examination (PATH International, 2025). Second, applicants must document at least 30 hours of equine-handling experience under a supervising instructor, complete a two-part skills checklist, complete 25 hours of experience working directly with individuals with disabilities, and complete cardiopulmonary resuscitation and first-aid certifications (PATH International, 2025). Finally, applicants must submit a 20-minute video demonstrating riding instruction and communication skills, provide evidence of at least 25 hours of supervised mounted therapeutic riding instruction under a CTRI-trained instructor, and take a CTRI certification exam (PATH International, 2025). Instructors oversee the session and coach riders as they perform new or acquired skills. The horse leader is responsible for leading the horse and attending to any of their needs. The side walker(s) walk alongside the rider while interacting and providing any necessary support to ensure that the rider remains safe while on and off the horse. Ultimately, clear and consistent communication between individuals is necessary to provide a safe and effective session for everyone involved.

Depending on the child's support needs, TR sessions at this facility normally begin with the rider assisting in the grooming and tacking process before participating in any mounted activity. The rider will complete this process under the direction of their instructor and with assistance from their side walkers (s). While grooming, riders use specific tools and brushes to groom horses and remove any dirt or debris before saddling. The grooming tools are used in a specific order on the horse and include what are called a curry, dandy, body, mane, tail curry, and hoof pick. After these tasks are completed, riders begin to tack up their horses. This is when the equipment is placed on the horse in preparation for riding. Tack equipment can vary according to horse, rider, and skill level. Riders normally assist with placing an underpad, saddle, girth, and reins on their horse, and adjust each piece accordingly to ensure a good fit. Because this process requires a specific step-by-step order, it encourages the rider to practice attending to various tasks, following directions. The grooming and tacking process can target many skills such as gross and fine motor coordination, cognitive thinking and processing, executive functioning, and social communication.

As soon as the grooming and tacking tasks are complete, the rider then puts on their helmet, with a side walker or instructor's assistance if needed. The rider then makes their way to the mounting block, and the instructor helps the child to get on the horse safely. They then move to the arena to begin riding activities. Activities in the arena can vary but often include fun interactive games connected with the performance of specific riding skills. For example, games such as red-light green-light, tic tac toe, and identifying and reaching for stuffed animals located in various places around the arena. Performing riding skills while participating in various games and activities can target a variety of skills, such as social communication/verbalization, fine/gross motor coordination, motor planning, task initiation, and command following. Riding a horse requires riders to utilize various parts of their body, such as muscles and joints, to make positional changes in response to the horse's movements, which may lead to improvements in skills such as motor coordination, organization, and planning in children with disabilities (Gabriels et al., 2012). Depending on the skill level of the rider, riders are often instructed to perform different riding skills, such as a two-point or posting riding position while the horse is at a walk/trot. These skills challenge the rider in maintaining balance and control of their body, ultimately requiring a higher level of gross motor coordination to perform. Gabriels et al. (2012) expressed the importance of demonstrating balance while on the horse, and the different movements of the horse encouraged the rider to continuously adjust their body weight to maintain postural control. Riders are encouraged to communicate with their horses throughout the session by giving them various verbal cues such as "walk on" when they are ready to go or "whoa" when they want to stop. This encourages the child to communicate with the horse and targets their ability to communicate effectively. After all riding activities were complete, all participants rode into the middle of the arena to thank their horses and volunteers before dismounting.

Autism Spectrum Disorder

A significant percentage of the research found on the effects of TR includes children with Autism Spectrum Disorder (ASD). Therefore, an in-depth review of the research regarding this population was conducted to determine whether any gaps were found in the literature. This literature review provides information on the impact of EAT on children with ASD.

ASD presents as a group of disorders that pose various challenges to children while functioning within their specific environments. These challenges result from deficits in social communication and interaction as well as cyclical patterns of interests, activities, and behavioral tendencies that affect their ability to adequately interact with peers and in social relationships (American Psychological Association, 2022). ASD can be diagnosed in children as early as 2 years of age; however, its onset is generally seen during preschool years (American Psychological Association, n.d.; National Institute of Mental Health, 2021). Children with ASD may face increased difficulties within their environment, which can be attributed to the societal norms and expectations placed on them by Western cultures. These children have unique forms of communication and behavior that may not always be accepted today and can prevent them from participating fully within their environment.

Animal-assisted therapy has become an effective method for improving the function of children with ASD (Chen et al., 2022; O'Haire, 2013). Several studies have evaluated the impact of TR in individuals with ASD. Research has shown that TR can improve behavioral deficits, such as lethargy, inappropriate speech, irritability, and hyperactivity, in children with ASD (Chen et al., 2022; Gabriels et al., 2015; O'Haire, 2013). These studies also found improvements in social communication after riding intervention (Chen et al., 2022; Gabriels et al., 2015; O'Haire, 2013). Improvements in social behavior and communication were associated with children's verbal and nonverbal interactions with horses, peers, and instructors, which helped them gain a better understanding of others (Chen et al., 2022). Not only have improvements been seen in behavior and social communication, but participation in TR has also been shown to have a significant influence on psychological well-being, self-efficacy, and self-esteem components in children with ASD, which ultimately positively impacts their overall quality of life in a positive way (Zhao et al., 2021).

One study examined the effects of equine-assisted activities in children with ASD combined with other comorbidities. Srinivasan et al. (2018) analyzed children with ASD and others that also had comorbidities such as Attention Deficit Hyperactivity Disorder (ADHD), hypersensitivity, or sensory integration disorder and concluded that equine-assisted therapy had promising positive effects on behavioral and sensory skills in these children, but evidence for social communication and motor skills was limited. The sensory Integration Frame of Reference emphasizes how sensory inputs, such as vestibular, proprioceptive, and tactile experiences, support adaptive responses and functional participation (Schaaf & Davies, 2010). During TR, a rider may receive rich vestibular and proprioceptive inputs due to the rhythmic, repetitive, and multidimensional movements provided by horses. Thus, it is unsurprising that we see preliminary evidence supporting sensory skills in children with ASD; however, more research is needed to understand the effects of TR specifically on sensory-related outcomes.

Research shows that numerous benefits have been found from TR for individuals with ASD; however, after a thorough review of the literature, most of the research explained previously focused on specific categories or challenges people with ASD often experience rather than providing a well-rounded approach with information on multiple deficit domains that can be present, such as behaviors, social skills, sensory processing, and independence during activities of daily living (i.e., activities performed on a daily basis, such as hygiene tasks, meal preparation, feeding oneself, etc.). Further, much of the research lacks high-quality designs (O'Haire, 2013). This lack of quality research should be addressed to provide more information on the role of TR in individuals with ASD who experience deficits that impact their overall health and wellbeing. Thus, the purpose of this research is to describe the perceived effect of TR on children with ASD using a survey that measured parents' and/or caregivers' perceptions of changes in behavioral, social, sensory, and confidence, as well as activities of daily living. We hypothesized that caregivers would perceive that TR would positively benefit all outcome areas.

Method

A quasi-experimental study design was utilized to determine whether the TR program had an impact on children with ASD. Quasi-experimental designs are research designs that evaluate the effects of an intervention but do not have a comparison group or where participants are not randomly assigned to groups. This study did not include a comparison group or a random assignment. The goal of this research was to evaluate the impact of TR on perceived changes in social, behavioral, and sensory components, along with independence in daily activities, since the child began participating in TR. Surveys were administered to parents/or caregivers of the participating riders, asking them to reflect on any perceived changes seen since their child first began TR. The surveys collected quantitative and qualitative data from January 6th to February 24th. We hypothesized that caregivers would perceive positive changes in all domains.

Participants

Parents/caregivers of children aged 4-12 years with ASD participated in this study. To be included in the study, parents must have a child who has participated in the TR program.

Setting

This study was completed at a non-profit organization in East Tennessee that provides equine-assisted experiences to individuals with disabilities or other support needs. This organization serves individuals aged 4-65 years and older, from many surrounding communities.

Procedures

All procedures involving human participants were performed in accordance with the ethical standards of the institutional and/or national research committee and the World Medical Association's Declaration of Helsinki, an international set of ethical principles that guides research with human participants, ensuring informed consent, risk minimization, independent review, and protection of participant rights and welfare (World Medical Association, 2024). The Institutional Review Board of Lincoln Memorial University determined that this research was exempt.

Through collaboration with the program director, potential participants were identified based on their children's ages and diagnoses. Once identified, the first author spoke with the caregivers of each child to explain the purpose of the study and provided recruitment flyers that included details on how to participate and a QR code to access the survey. An email was sent to each caregiver as an additional method of recruitment. The email consisted of a study description and provided a link to access the survey. Those who scanned the flyer QR code or clicked the link on email were directed to the survey and asked to provide consent before completion.

Measures

A survey design process was completed to increase survey validity by carefully developing the survey items through a multi-step process (Artino et al., 2014). First, a literature review was conducted to determine gaps and assess the outcome measures utilized in the existing research. Second, interviews were completed with experienced TR instructors and parents of children with ASD who were enrolled in the TR program at a non-profit organization where the research took place. These interviews were used to identify the potential outcomes observed after engaging in TR. Interview responses were transcribed verbatim and analyzed using content analysis with both authors to determine themes and potential outcome domains to be included in the survey. Next, an additional literature review was conducted, and themes from the literature review were mapped onto the findings from the interviews to determine potential survey topics. Survey items were then developed, feedback was collected through expert validation (i.e., review and feedback from TR instructors, parents, occupational therapists, and researchers with expertise in survey development), and changes to survey items were made accordingly.

A Qualtrics survey was administered to the parents of the children who participated in the TR program. This measure was used to assess parental perceptions of change in their children and specifically targeted behavioral (i.e., ability to follow directions, focus on tasks, regulate or co-regulate emotions, and one's self-control), sensory (i.e., frequency of repetitive movements and changes to sensitivity to sights, sounds, textures, or other sensory input), social (i.e., ability to show empathy towards others, communication, recognition and response to social cues, initiation and response to conversations), relationships (i.e., satisfaction with friendships and family relationships, number of social interactions, and satisfaction with social interactions), confidence (i.e., at school, home, in the community, with friends, when completing daily activities, when trying new things, and with one's self-image), and activities of daily living categories (i.e., toileting, dressing, bathing, caring for pets, preparing meals, feeding self, chores, personal hygiene, play, and sleep). The survey used a five-point Likert scale ranging from a lot better, a little better, no change, a little worse, to a lot worse.

One additional open-ended question was incorporated to gather information on any other changes observed in addition to the main survey categories.

Statistical Analysis

Quantitative data gathered from the survey were analyzed using descriptive statistics, specifically, the frequency and percentage of each survey response by category. Descriptive statistics were used to provide a clear summary of the key findings and identify patterns or trends related to perceived outcomes. Qualitative data were analyzed using content analysis to identify overarching themes from the parent/caregivers' open-ended responses. Specifically, the first author prepared the data, both authors reviewed the open-ended responses multiple times, breaking the text into units of meaning and determining initial "codes" (Elo & Kyngas, 2007). Next, analysts reviewed the codes and grouped them to form subcategories for reporting (Elo & Kyngas, 2007).

Results

Seventeen parents and/or caregivers of children aged 5–12 years participated in this study. Only one participant completed half of the survey. Their data were maintained, but demographic characteristics were not reported for this individual. Ten of those children were male, 5 female, and one preferred not to specify their gender. Seven children participated in TR for less than a year, eight for 1-2 to years, and one for five to six years. Respondents shared that occupational therapy, speech therapy, special education, and applied behavior analysis were the most common additional services their children received or were receiving, while the least common were physical therapy and mental health counseling. Parents primarily reported that their children had co-occurring conditions of sensory processing disorder, attention deficit hyperactivity disorder, and intellectual disability, whereas anxiety, sleep disorder, and obsessive-compulsive disorder were selected less frequently (Table 1).

Table 1
Demographic Characteristics

Demographic	Frequency	Percent
Age		
5	2	12.5
6	4	25.0
7	1	6.3
9	5	31.3
11	3	18.8
12	1	6.3
Gender		
Male	10	58.8
Female	5	29.4
Prefer not to say	1	5.9
Years of participation in Therapeutic Riding		
Less than 1 year	7	41.2
1-2 years	8	47.1
5-6 years	1	5.9
Other services received and/or receiving		
Occupational Therapy	10	58.8
Physical Therapy	3	17.6
Speech Therapy	11	64.7
Mental health/counseling	4	23.5
Applied Behavioral Analysis	7	41.2
Special Education	9	52.9
No other services received	2	11.8
Diagnoses		
Intellectual disability	5	29.4
Attention Deficit Hyperactivity Disorder	6	35.3
Sensory Processing Disorder	9	52.9
Sleep Disorder	1	5.9
Anxiety	4	23.5
Depression	1	5.9
Obsessive Compulsive Disorder	1	5.9

Perceived Skill Development from Therapeutic Riding

The behavioral category showed the greatest percentage of improvement across all areas of the survey. Most parents reported positive changes in their child's ability to follow directions, attention, and focus on tasks, emotional regulation, and the ability to co-regulate (88%), while 82% reported positive changes in their child's self-control. These results suggest that parents observed notable improvements in their children's behavioral tendencies as a result of the TR. After assessing responses in the sensory processing category, 76.5% of respondents indicated positive changes in their children's sensitivity to sights, sounds, textures, or other sensory inputs after the TR intervention. However, only 41.1% of the participants reported a positive change in the frequency of repetitive movements (e.g., hand flapping, rocking, and spinning). Notably, none of the parents reported a decrease in behavioral or sensory outcomes (Table 2 & Figure 1).

Table 2

Behavior and Sensory Regulation skill development

Skills	Frequency	Percent
Behavior		
Ability to follow directions		
A lot better	8	47.1
A little better	7	41.2
No change	2	11.8
Attention or focus on tasks		
A lot better	4	23.5
A little better	11	64.7
No change	2	11.8
Emotional regulation		
A lot better	7	41.2
A little better	8	47.1
No change	2	11.8
Ability to co-regulate		
A lot better	8	47.1
A little better	7	41.2
No change	2	11.8
Self-control		
A lot better	6	35.3
A little better	8	47.1
No change	3	17.6
Sensory		
Frequency of repetitive movements		
A lot better	4	23.5
A little better	3	17.6
No change	10	58.8
Changes to sensitivity to sights, sounds, textures, and other sensory input		
A lot better	5	29.4
A little better	8	47.1

No change	4	23.5

Figure 1
Improvements in Behavior and Sensory Regulation Skill Development

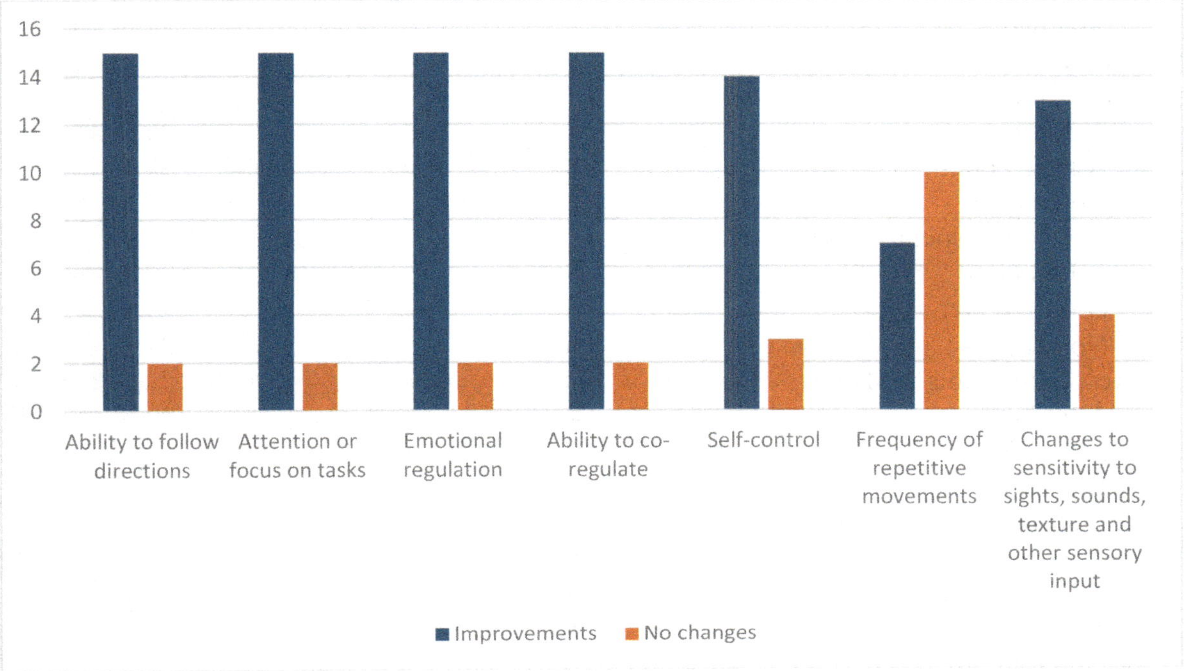

Regarding social skills, 82.4% of the parents who participated in the study reported improvement in their child's ability to communicate with peers, friends, and family members, while 70.6% reported that their child's ability to respond to social cues was better after TR intervention. While responses targeting the ability to show empathy, recognize social cues, initiate conversations, and respond to others/maintain conversations were not as noteworthy, over half of the sample indicated that they observed improvements in these areas. Many participants indicated that their children's satisfaction with family relationships and social interactions had increased (70.6%), while satisfaction with friendships did not rank as high as there were only 47% of respondents who reported improvement. Since the beginning of the TR services, 64.7% indicated that there had been an increase in the number of social interactions with others. Notably, none of the parents reported a decrease in their child's social skills or relational outcomes. Overall, these results indicate that TR has a positive impact on children's ability to communicate effectively with others around them (Table 3 & Figure 2).

Table 3
Social Skills and Relationships development

Skills	Frequency	Percent
Social		
Ability to show empathy towards others		
A lot better	3	17.6
A little better	6	35.3
No change	5	29.4

Communication with peers and friends		
A lot better	6	35.3
A little better	8	47.1
No change	2	11.8
Communication with family members		
A lot better	6	35.3
A little better	8	47.1
No change	2	11.8
Recognizing social cues		
A lot better	5	29.4
A little better	5	29.4
No change	6	35.3
Responding to social cues		
A lot better	4	23.5
A little better	8	47.1
No change	4	23 5
Initiating conversations		
A lot better	6	35.3
A little better	5	29.4
No change	5	29.4
Responding to others/maintaining conversations		
A lot better	4	23.5
A little better	6	35.3
Relationships		
Satisfaction with friendships		
A lot better	3	17.6
A little better	5	29.4
No change	8	47.1
Satisfaction with family relationships		
A lot better	6	35.3
A little better	6	35.3
No change	4	23.5
Number of social interactions		
A lot better	4	23.5
A little better	7	41.2
No change	5	29.4
Satisfaction with social interactions		
A lot better	4	23.5

A little better	8	47.1
No change	4	23.5

Figure 2
Improvements in Social Skills and Relationships Outcomes

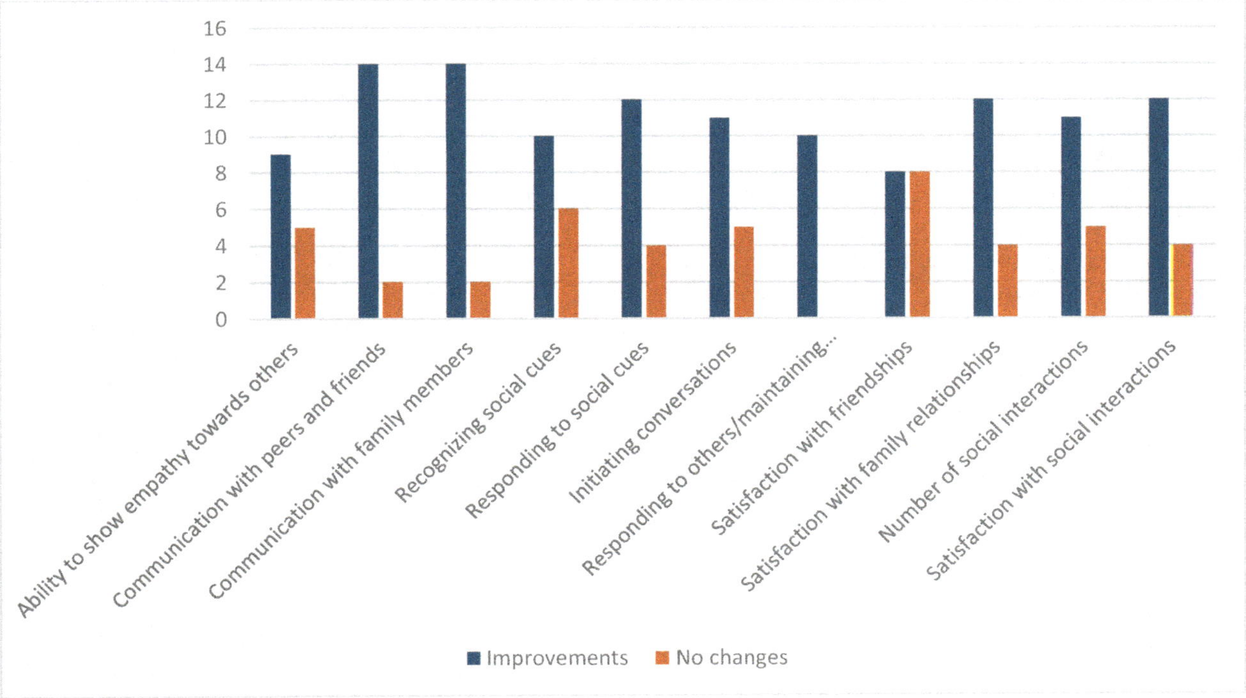

Several positive outcomes were noted in the confidence category. Over half of the parents who completed the survey reported an increase in their children's confidence in all seven confidence areas. The most impactful positive changes were noted in the children's confidence at home (76.4%) and when trying new things (82.4%). Many participants indicated that they had seen an increase in confidence while at school, in the community, or when completing daily activities. The percentages of least improvement were observed in the areas of self-image (64.7%) and confidence with friends (58.8%). Again, no parents reported a decrease in confidence (Table 4 & Figure 3).

Table 4
Confidence

Confidence	Frequency	Percent
At school		
A lot better	5	29.4
A little better	7	41.2
No change	4	23.5
At home		
A lot better	10	58.8
A little better	3	17.6
No change	3	17.6
In the community		

A lot better	7	41.2
A little better	5	29.4
No change	4	23.5
With friends		
A lot better	6	35.3
A little better	4	23.5
No change	6	35.3
When completing daily activities		
A lot better	8	47.1
A little better	4	23.5
No change	4	23.5
When trying new things		
A lot better	7	41.2
A little better	7	41.2
No change	2	11.8
With self-image		
A lot better	6	35.3
A little better	5	29.4
No change	5	29.4

Figure 3
Improvements in Confidence

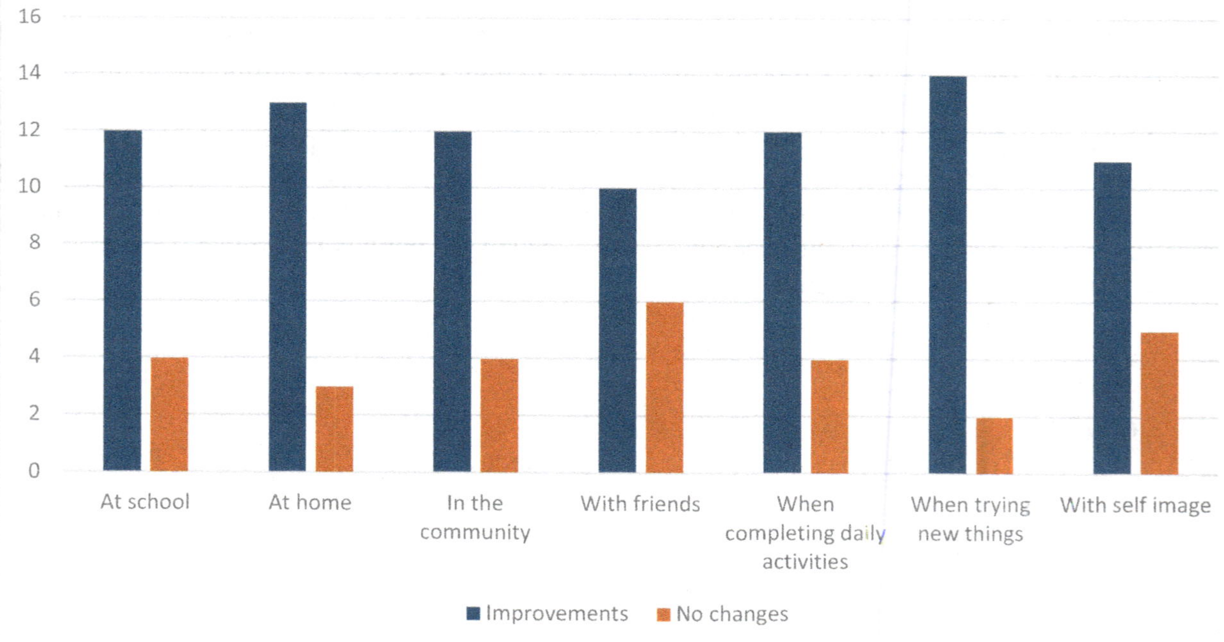

When breaking down the results of independence in daily activities, 64.7% of the participants reported that they had seen positive changes in their children's play, sleep, and dressing activities. Slightly over half of the participants observed improvements while caring for pets, bathing, and during chore completion. However, only some improvements were noted in personal hygiene, self-feeding, and meal preparation. While improvements were observed, the amount of positive change was much less when compared to other daily activity areas. However, none of the parents reported worsening of their child's skills after starting TR. (Table 5 & Figure 4)

Table 5
Independence in daily activities skill development

Independence in daily activities	Frequency	Percent
Toileting		
A lot better	5	29.4
A little better	4	23.5
No change	7	41.2
Dressing		
A lot better	7	41.2
A little better	4	23.5
No change	5	29.4
Bathing		
A lot better	5	29.4
A little better	4	23.5
No change	7	41.2
Caring for pets		
A lot better	6	35.3
A little better	4	23.5
No change	6	35.3
Preparing meals		
A lot better	1	5.9
A little better	3	17.6
No change	12	70.6
Feeding self		
A lot better	2	11.8
A little better	4	23.5
No change	10	58.8
Chores		
A lot better	2	11.8
A little better	8	47.1
No change	6	35.3
Personal hygiene		
A lot better	2	11.8
A little better	5	29.4
No change	9	52.9
Play		
A lot better	6	35.3
A little better	5	29.4

No change	5	29.4
Sleep		
A lot better	5	29.4
A little better	6	35.3
No change	5	29.4

Figure 4
Improvements in Activities of Daily Living

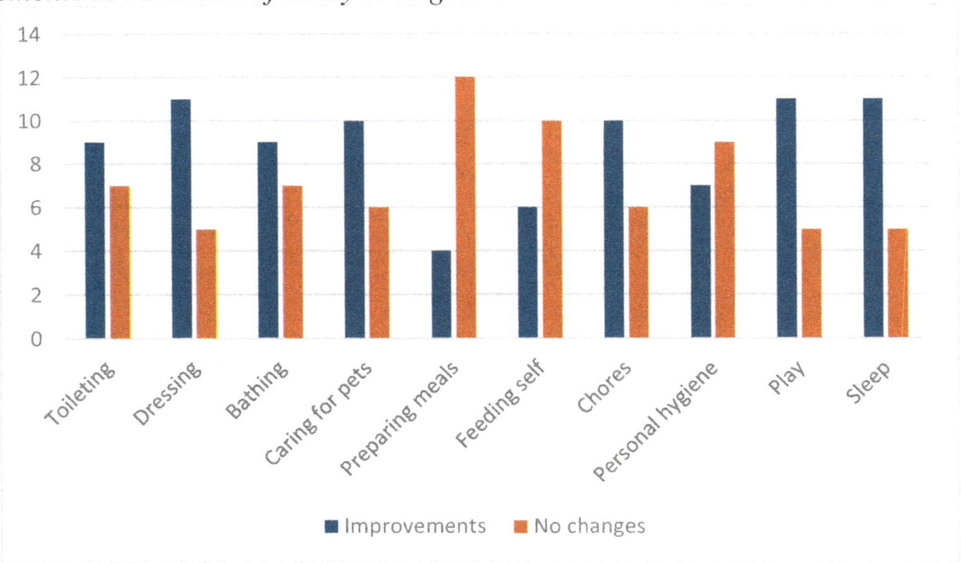

Qualitative Findings

Four overarching themes were identified from the parent/caregivers' open-ended responses. *Confidence* is presented as a common theme. Participants reported that they observed an improvement in confidence and self-efficacy among their children in various respects. For example, one respondent noted, "I have never seen [child's name] so confident as when he was riding Gordy!" While another stated, "My son enjoys coming to [the organization name], he joyfully participates in the activities, and he has overcome his fear of horses. As a result, he is less afraid of other animals." *Happiness and excitement* were the second most common themes. Several parents reported that they noticed an increased level of joy and satisfaction with their child since they started riding. One parent exclaimed, "He woke up every Wednesday and the first words out of his mouth were always, "I get to see Gordy today!" The third most common theme presented from the data was *attention* and *following directions*. Parents noted impactful positive changes in their children's directed attention and ability to follow commands as a result of TR. One parent wrote, "The increase in attention span has been incredible to watch." The final theme was *motivation and willingness to try new things*. Participants indicated that they witnessed their child display an increased level of determination and eagerness to participate in new activities. As one parent reports, "We have never seen [the child's name] try so hard, listen so well, and be so focused on getting it right when it came to riding. In other previous activities, he did not like it and made zero effort – so this was a huge change for us."

Discussion

The purpose of this research was to describe the perceived effect of TR on children with ASD across behavioral, social, sensory, confidence, and activities of daily living outcomes. We hypothesized that caregivers would perceive the children as demonstrating positive outcomes across all measured areas following participation in TR. The findings of this study suggest that, through the lens of parents and caregivers, TR may be used as an effective intervention to improve outcomes in children with ASD. Positive changes were identified in all categories of the study, although the most notable impacts were observed in domains such as behavior, social communication, and confidence. Other studies have mirrored similar findings. For example, multiple studies have found improvements in directed attention, sensory sensitivity, social motivation, social communication, and reactions to sensory input after TR (Bass et al., 2009; O'Haire, 2013; Ward et al., 2013).

During TR sessions, children were encouraged to follow purposeful routines and were instructed to perform specific tasks while on and off the horse. These activities could potentially foster the children's ability to follow directions and attend to tasks that may have otherwise been unsuccessful without the use of equine animals. Participation in TR can also provide opportunities for children to receive an experience filled with multiple sensory components (Ward et al., 2013), in alignment with the Sensory Integration Frame of Reference (Schaaf & Davies, 2010). As a result, the majority of parents and/or caregivers indicated that they observed sensory improvements in their child with less sensitivity to sensory input such as sights, sounds, and textures. However, restricted repetitive movements were less commonly reported as changing in this study. This may be due to restricted repetitive movements being the core feature of ASD. These movements may be more neurologically driven, rather than skill- or motivation-based behaviors that can be readily changed by TR.

Interaction with horses has been shown to trigger physiological, psychological, and social responses in children (O'Haire, 2013; Ward et al., 2013). During TR, instructors and volunteers provide instructions to riders and encourage them to give commands and communicate with their horses, which could influence the improvements demonstrated in social communication (Gabriels et al., 2015). However, because most social communication occurs between the rider and the horse or between the rider and the instructors, it is unsurprising that caregivers in this study perceived social interactions with peers or friends as less common. It is possible that the use of the horse increases motivation in children with ASD to effectively interact and communicate with their equine partner and those around them, which may encourage the transfer of these skills into other areas of daily life (Bandura, 1977; Deci & Ryan, 1985). The study findings also show that participation in TR can lead to greater feelings of confidence in children with ASD, which could ultimately encourage participation in new skills and activities outside one's normal routine and existing environment. Again, given the nature of TR and the limited opportunities for developing communication and interaction skills with peers, it is unsurprising that caregivers perceived confidence with peers or friends as less common after engaging in TR in this study.

Strengths, Limitations, and Future Research

One strength is that the study encompassed a survey-based approach with both quantitative and qualitative data to obtain a well-rounded understanding of parental perceptions of the impact of TR. However, this study has several limitations. There are several limitations related to the purposive sample used in this study. First, all the children had received or were currently receiving other therapy-related services while participating in the TR, such as occupational, speech, or applied behavior analysis, which creates potential confounding variables. Additionally, each participant's child may have received multiple and varying sessions of TR across instructors, which could influence the outcomes. We also did not collect data on ASD severity, which could have influenced the effects of TR on various outcomes. Lastly, survey respondents were not given the option to select whether their child had any prior history of repetitive movements, which could have affected the results of that particular survey item.

There are also inherent limitations to the study design. Specifically, the research employed a quasi-experimental design and there were no baseline data collected or a control group for comparison; thus, it is difficult to distinguish whether all positive changes were specifically correlated with participation in TR as other factors could have contributed to the observed improvements. The survey was also developed by researchers, lacking established reliability or validity, and was based on parents' subjective views, which could be seen as another limitation. Additional variation in TR sessions and instructors may have further influenced the results. Additional research is needed to provide more information on the benefits of TR in this population. Future research should utilize randomized controlled trial designs with pre-test–post-test data collection points using reliable and valid measures across the following outcomes: behavior, social skills, and confidence. Research could also be improved by including TR observations to ensure fidelity and standardization across TR sessions and interventions.

Conclusion

Parents and/or caregivers of children with ASD observed many positive changes in their children after participating in TR, with the most impactful changes identified in behavior, social communication, and confidence. There were no negative responses or instances in which any deficits worsened as a result of participation. Therefore, the findings suggest that TR may be used as an efficacious activity to improve deficits and promote skill development in behavioral, social, sensory, confidence, and activities of daily living factors in children with ASD. This improvement is ultimately achieved through the development of riding skills and by utilizing the power of equine animals to provide therapeutic experiences. Although positive outcomes were identified, additional research is needed to further support these findings.

Data Availability Statement: Data available on request from the corresponding author.

Conflict of Interest: The authors declare no conflicts of interest.

Funds: No funds were received.

Author Contributions: Conceptualization, AS, EKS; methodology, AS, EKS; formal analysis, AS, EKS; investigation, AS, EKS; resources, AS, EKS; data curation, AS, EKS; writing-original draft preparation, AS, EKS.; writing-review and editing, AS, EKS; project administration, AS, EKS. All the authors have read and agreed to the published version of the manuscript.

Statement: The study was conducted in accordance with the guidelines of the Declaration of Helsinki and approved by the Institutional Review Board of Lincoln Memorial University (IRB #2024/02/35).

References

American Psychological Association. (n.d.). autism spectrum disorder (ASD). *In APA dictionary of psychology*. Retrieved January 30, 2025, from https://dictionary.apa.org/autism-spectrum-disorder?

American Psychiatric Association. (2022). *Diagnostic and statistical manual of mental disorders* (5th ed., text rev.). https://doi.org/10.1176/appi.books.9780890425787

Artino, A. R., La Rochelle, J. S., Dezee, K. J., & Gehlbach, H. (2014). Developing questionnaires for educational research: AMEE Guide No. 87. *Medical Teacher*, *36*(6), 463–474. https://doi.org/10.3109/0142159X.2014.889814

Bandura, A. (1977). Self-efficacy: Toward a unifying theory of behavioral change. *Psychological Review, 84(2),* 191–215. https://doi.org/10.1037/0033-295X.84.2.191

Bass, M. M., Duchowny, C. A., & Llabre, M. M. (2009). The effect of therapeutic horseback riding on social functioning in children with autism. *Journal of Autism and Developmental Disorders*, *39*, 1261–1267. https://doi.org/10.1007/s10803-009-0734-3

Chen, S., Zhang, Y., Zhao, M., Du, X., Wang, Y., & Liu, X. (2022). Effects of therapeutic horseback-riding program on social and communication skills in children with autism spectrum disorder: A systematic review and meta-analysis. *International Journal of Environmental Research and Public Health*, *19*(21), 1–13. https://doi.org/10.3390/ijerph192114449

Deci, E. L., & Ryan, R. M. (1985). *Intrinsic motivation and self-determination in human behavior*. Plenum Press. https://doi.org/10.1007/978-1-4899-2271-7

Elo, S., & Kyngäs, H. (2008). The qualitative content analysis process. *Journal of Advanced Nursing, 62(1),* 107–115. https://doi.org/10.1111/j.1365-2648.2007.04569.x

Gabriels, R. L., Agnew, J. A., Holt, K. D., Shoffner, A., Zhaoxing, P., Ruzzano, S., Clayton, G. H., & Mesibov, G. (2012). Pilot study measuring the effects of therapeutic horseback riding on school-age children and adolescents with autism spectrum disorders. *Research in Autism Spectrum Disorders*, *6*(2), 578–588. https://doi.org/10.1016/j.rasd.2011.09.007

Gabriels, R. L., Pan, Z., Dechant, B., Agnew, J. A., Brim, N., & Mesibov, G. (2015). Randomized controlled trial of therapeutic horseback riding in children and adolescents with autism spectrum disorder. *Journal of the American Academy of Child and Adolescent Psychiatry*, *54*(7), 541–549. https://doi.org/10.1016/j.jaac.2015.04.007

National Institute of Mental Health. Early Intervention for Autism. Retrieved February 1, 2025 from https://www.nichd.nih.gov/health/topics/autism/conditioninfo/treatments/early-intervention

O'Haire, (2013). Animal-Assisted Intervention for Autism Spectrum Disorder: A systematic literature review. *Journal of Autism and Developmental Disorders, 43,* 1606-1622. DOI: 10.1007/s10803-012-1707-5.

PATH International (2022). *PATH International equine-assisted services definitions.* https://pathintl.org/wp-content/uploads/2022/03/PATH-Intl-Services-Terminology-Centers.pdf.

PATH (2025). *Summary of updates to the PATH International CTRI application requirements and exam administration policies for 2025.* (n.d.). https://pathintl.org/wp-content/uploads/2024/10/Summary-of-2025-CTRI-Updates.pdf

Schaaf, R.C. & Davies, P.L., (2010). Evolution of the sensory integration frame of reference. *Department of Occupational Therapy Faculty Papers.* Paper 18. https://jdc.jefferson.edu/otfp/18.

Srinivasan, S. M., Cavagnino, D. T., & Bhat, A. N. (2018). Effects of equine therapy on individuals with autism spectrum disorder: A2 systematic review. *Review Journal of Autism and Developmental Disorders*, *5*(2), 156–175. https://doi.org/10.1007/s40489-018-0130-z

Ward, S. C., Whalon, K., Rusnak, K., Wendell, K., & Paschall, N. (2013). The association between therapeutic horseback riding and the social communication and sensory reactions of children with autism. *Journal of Autism and Developmental Disorders*, *43*, 2190–2198. https://doi.org/10.1007/s10803-013-1773-3

World Medical Association, (2024). *WMA Declaration of Helsinki – Ethical Principles for Medical Research Involving Human Participants.* https://www.wma.net/policies-post/wma-declaration-of-helsinki/.

Zhao, M., Chen, S., You, Y., Wang, Y., & Zhang, Y. (2021). Effects of a therapeutic horseback riding program on social interaction and communication in children with autism. *International Journal of Environmental Research and Public Health*, *18*(5), 1–11. https://doi.org/10.3390/ijerph18052656

Effects of Implementing Music Therapy Concurrently with Adaptive Riding: A Pilot Study

Aviva Vincent, LSW, Ass. Professor[1], *Emmet Evans, BS*[1], *Amanda Wilson-Blevens, MSW, LSW*[1], *Claudia Litherland, BA*[1], *Rylan Roberts, BS*[2], *Dustin Hildenbran, MS*[2], *CDCA, and Linda Quinn, PhD*[2]

[1]School of Social Work, College of Health, Cleveland State University, Ohio, USA, [2]Math and Statistics, College of Arts and Sciences Cleveland State University, Ohio, USA

This observational pilot study evaluated the effect of integrating music therapy within adaptive riding lessons for youth with cognitive and social-emotional disabilities. The study design included an intervention group (adaptive riding with music therapy) and a control group (adaptive riding). To obtain observational data, researchers used the Gainesville Riding through Equine Assisted Therapy Scale (GREAT Scale) and the Global Assessment Scale (GAS) for Equine Assisted Services (EAS). Over two eight-week sessions, participants (n=16) were evaluated regarding focus, horsemanship skill/posture, self-regulation, enjoyment, and heart rate. To collect heart rate data, Polar Verity Sense heart rate monitors were placed on Certified Therapeutic Riding Instructors (CTRI), volunteers, and students. To collect equine heart rate data, Polar Verity Sense equine heart rate sensors were placed on the horses. Descriptive statistics and normality for each group was assessed via visual analysis, Q-Q plots, and Shapiro-Wilk test. The GREAT Scale data were compared using a two-sample t-test if both groups were normal, and Welch or Mann-Whitney as appropriate when distribution was not normal. The GAS scores were averaged for use in a multivariate regression and multiple linear regression. The average and maximum heart rate per session were analyzed by frequency and distribution. Improvements in focus, horsemanship/posture, self-regulation, and enjoyment were evident in both groups, however, improvements were significantly higher for the intervention group in horsemanship/posture. No statistically significant differences were found in heart rate data. The addition of music therapy to adaptive riding lessons may offer increased benefits in areas of focus, horsemanship/posture, self-regulation, and enjoyment.

Keywords: adaptive riding, equine-assisted services, heart rate, horse and human interaction, music therapy

All correspondence and requests for reprints should be addressed to: Aviva Vincent, at a.l.vincent@csuohio.edu or Cleveland State University 2121 Euclid Ave., Cleveland, Ohio, 44115, USA.

Handling Editor: Alexandra Stergiou
Reviewers: Lisa Baugh, Laura Dennis, Lorenzo Lucarelli

Volume 25, Number 1, January 2026

DOI: https://doi.org/10.58862/BPEJ3021

Both music and horses have significantly influenced human history. Music has long served purposes of communication and emotional expression with the oldest musical instruments dating back 40,000 years (Hallberg, 2018; Schulkin & Raglan, 2014). The benefits of horseback riding were noted by Hippocrates with horses being depicted in cave paintings dating back 25,000 years (Pruvost et al., 2011; Schulkin & Raglan, 2014).

Adaptive riding is a modality of partnering with horses that adapts horseback riding techniques for individuals with "disabilities, health issues, and other major life challenges" (Wood et al., 2021, p. 89). Music therapy involves using music interventions to meet therapeutic goals (American Music Therapy Association [AMTA], n.d.). Within the United States, both practices are guided by professional associations which ensure the credentialing of professionals. Relevant to this study, Music Therapists were credentialed by AMTA for music therapy and the Professional Association for Therapeutic Horsemanship International (PATH International) for adaptive riding (AMTA, 2024; PATH International, 2024).

There can be a variety of purposes for utilizing music therapy and adaptive riding, independently and concurrently, such as helping individuals build and expand foundational life skills, practice clear communication, improve gross motor function, regulate behavior and emotions, and achieve personal goals (Bruscia, 1988; Kaplan & Steele 2005). Furthermore, horses and humans may experience co-regulation between evidenced by heart rate variability and synchronization (von Lewinski et al., 2013; Gehrke et al., 2011). Even though both horses and music have separate influences human history and culture, and despite demonstrated interest from practitioners of both fields to combine modalities, there remains a dearth of empirical research on the impact of implementing music therapy concurrently with adaptive riding (Rossi, 2018).

This study investigated the addition of music therapy to create an opportunity for experiencing different sensory, gross motor, and social skills in adaptive horseback riding compared to traditional adaptive riding without music therapy. It has been postulated that beat and rhythm of music may have a direct influence on physical and psychological systems in humans (Jabborova & Mohirabonu, 2023) and horses (Gueguen et. al., 2023), which may be a physiological indicator of self-regulation, co-regulation, and enjoyment. Thus, the inclusion of music therapy in adaptive horseback riding may affect the emotional state of the horse, which in turn could positively influence the experience of the student rider (Ferlazzo et. al., 2024).

Music therapy interventions require participants in adaptive riding to engage in receptivity, improvisation, performance, and entrainment (Rossi, 2018). Receptive listening is demonstrated through the act of attending to sounds and music, such as that of the horse's hoofbeats, one's own breathing, or to the instructions of the Certified Therapeutic Riding Instructor (CTRI), volunteer side-walker, or Music Therapist. Improvisation is demonstrated by the student or Music Therapist through use of instrumental or vocal music spontaneously. Performance occurs when students explore by demonstrating a movement learned via CTRI instruction, music therapy intervention, or observation of other riders. Entrainment is the action of allowing for body movement (of the rider) to follow the rhythm provided by an outside source (the horse), potentially without conscious thought. Taken together, receptivity, improvisation, performance, and entrainment provide opportunities for engaging in complex instructions. For example, a singing and instrumental song may be used to teach the skills of "walk on" and "whoa" (stop). By singing a "Walk On" song, the Music Therapist reiterates when to go/stop and provides directions for how to perform the necessary movements, cueing riders to anticipate and perform the behavior, process, and skill. The theoretical framework of Rhythmic Entrainment was the dominant foundation of this pilot study. Rhythmic Entrainment is postulated to affect "heartbeat, pain reduction, muscle relaxation, and music-mediated imagery" among participants in music therapy (Bharathi et al., 2019, p. 182). The auditory and proprioceptive senses are activated and engaged during music therapy, which may support a positive psychosomatic experience (Bruscia, 1988).

Developmental Theory, and Behavioral Theories are relevant to the cognitive process that may be underlying the intervention. Developmental Theory provides a method of predicting and describing human-horse-music interactions. Behavior Theory may support the hypothesis that adding music to adaptive riding lessons may create an environment of positive reinforcement for learning the skills, trying new and challenging tasks, and incentivizing engagement as these are core tenets of the theoretical framework (Bandura et al., 1977). Reybrouch and colleagues (2021) postulated that humans may be hard-wired to connect through music, as music is often used as an activating and regulating tool (Bonassi et al., 2023). Based on the theoretical foundation of the Developmental and Behavioral theories, music therapy may be particularly effective for youth as their brains are in formative stages of development and thus may be more primed to learn from music.

A scoping review (Paré & Kitsiou, 2017) was conducted using the Education Resource Information Center database (*ERIC)*. All accessible journals were included with no specific date range set, and only peer-reviewed and full-text available sources were collected. Using the keywords "adaptive horse* riding" and "music therapy*", the search yielded six hits with no duplicates. Two studies were relevant in that they explored the impact of music therapy and adaptive riding. However, the population in both studies were adults post-stroke.

Bunketorp-Käll and colleagues (2019) compared a group in adaptive riding to a group with music therapy, however, the interventions were not implemented concurrently. The researchers found that adaptive riding produced "immediate gains in gait and functional task performance in the late phase after stroke, whereas the effectiveness of [music therapy] is less clear" (p. 483). Rossi (2018) also wrote on this topic, however that paper did not appear in the search as it was an unpublished master's thesis. Gonzalez-Hoelling and colleagues (2024) published a systematic review of interventions for individuals' post-stroke, which concluded that "[g]iven the heterogeneity of the interventions, the interventions in control groups, the varied durations, and the different outcome measures…care should be taken in interpreting and generalizing findings" (p. 1).

The pilot study reported here investigated the effects of combining music therapy with adaptive riding (MT/AR) compared to traditional adaptive riding (AR). Specifically, it examined whether an intervention of integrating music therapy into adaptive riding enhanced focus, enjoyment, horsemanship skills, and self-regulation compared to the control of adaptive riding without music therapy. The study also explored differences in heart rate for riders, volunteers, and horses as a possible physiological indicator of self- regulation, co-regulation, and enjoyment. The research question was: Does music therapy increase students' focus/time on task, postural alignment, self-regulation, and overall enjoyment during adaptive riding lessons with music therapy, compared to adaptive riding lessons without music therapy?

Researchers hypothesized that students in the intervention group (MT/AR) would demonstrate improvements across all factors examined compared to those in the control group (AR).

Method

All procedures performed in this pilot study involving human and animal participants were done following the ethical standards of the Institutional Review Board at the Cleveland State University where the principal investigator was employed. Approved in 2019, (IRB-FY2024-24) the study was paused during COVID-19, then restarted in 2022, and concluded in 2024. The Principal Investigator (PI) recruited and trained five Research Assistants (RA) to support data collection; n = 4 Master of Social Work students, n = 1 Bachelor of Social Work student. Two RAs observed each session.

Design

Research for this study was conducted at a premiere PATH International accredited center offering Equine Assisted Services (EAS) mounted and unmounted equine activities in a Midwest suburb. The Center offered adaptive riding lessons to school groups in grades 4-12. Up to twenty students attended per session as part of their school day. The schoolteacher could request specific skills as a focus of the session, such as transitioning between activities or practicing appropriate interpersonal skills. Up to ten students were able ride at a time, with five in each of the two arenas.

Participants

Students who participated in adaptive riding as part of their school day were recruited and consented into the study by the PI. The students were randomly assigned to the control or intervention group. The division of groups relied on horse and volunteer availability, but once established, was consistent through the eight-week session.

The horses were all owned by the Center. While the horses' tenure with the Center varied from months to years, they experienced the same training, care, and work program. Each horse was consistently monitored for their physical and cognitive/emotional wellbeing by the Equine Director. All horses included in this study were exposed to music therapy prior to engaging in the music therapy intervention to ensure their comfort and welfare. If any horse demonstrated an aversion to adaptive riding, music therapy, or any part thereof, they were excused from the lesson as appropriate (personal communication, Fieldstone Farm, 2024).

Intervention

The Center provides programming, such as adaptive riding, in eight-week sessions. The study was conducted over the course of two sessions. For the intervention study, each week, two lessons were run concurrently in side-by-side arenas divided by a thick fabric divider. Approximately five students took part in each lesson. One lesson provided traditional adaptive riding without the addition of music therapy (control, AR), while the other lesson provided the addition of music therapy (intervention, MT/AR).

In this pilot study, equine welfare was established by partnering with a premiere accredited PATH International facility where high standards of practice are explicit when considering the role of the equine in any intervention. The eight weekly sessions generally included teaching the skills of: 1. Whoa and go, 2. Steering, 3. Steering with obstacles, 4. Steering with manipulatives (e.g., rings, bean bags), 5. Adding in trot, 6. Steering and Trot, 7. Trail Walk/outside, 8. Wrap up and student's choice. Within each skill, students could have been on or off lead, have had the support of a side-walker, and were encouraged to use verbal and physical aids. At any time, the CTRI could have modified the activities as needed based on the horse, student, and environment. During the intervention (MT/AR), the Music Therapist and CTRI co-facilitated the lesson. The Music Therapist integrated a variety of instruments, including voice, guitar, drum, and recorded music. Using songs, music, and sounds from instruments, students were guided through activities that engaged complex fine and gross motor movements, such as grooming, mounting, steering, stopping, and using manipulatives (i.e., bean bags, rings, etc.) while on horseback.

Outcome Measures

The first session tested all measures except for heart rate; the second session included all measures, including heart rate. The purpose was to slowly introduce measures over time to ensure a consistent process, data collection, and fidelity.

GREAT Scale

In the first session of eight weeks, the RAs observed and collected data using the Gainesville Riding through Equine Assisted Therapy, or GREAT Scale (Frease, 1996). The GREAT Scale measures postural control and balance through five specific areas of evaluation. Posture reflects the physical alignment of the student's body in the saddle. Horsemanship reflects the student's use of aids (i.e., hands/fingers on reins, heel position) while riding. The two constructs are grouped because there is an explicit connection between body position and the effective use of aids. The GREAT Scale was the only known scale identified that systematically assessed posture without the need for subjective knowledge of horseback riding. There are no published, accessible psychometric properties about the GREAT scale (Thompson et al., 2014).

Global Assessment Scale

The Global Assessment Scale (GAS) for Equine-Assisted Services (EAS) was used to measure the degree of attainment of the four goals, focus, enjoyment, horsemanship skills/posture, and self-regulation. for each individual rider (Smith Hybels, 2015). Because there were no known psychometric properties, the GAS was used as a pseudo-standardized measure by the RAs and instructors to rank student progress on individual and group goals; 1 for minimal progress and 5 for profound positive progress.

Heart Rate

The second eight-week session included the collection of heart rate data in addition to the GAS and GREAT Scale. The heart rate data was gathered using Polar Verity Sensors, Kempele, Finland (Polar, 2022) on CTRIs, volunteers, and riders Polar, 2022Merrigan et al., 2023; Navalta et al, 2023), and the Polar Verity Sense equine heart monitor on each horse (Bitschnaul, 2014; Kapteijn et al., 2022). While baseline and average vary by species, heart rate was intended to be used as an indicator of stress and engagement. Changes in heart rate may evidence co-regulation between horse and human (von Lewinski et al., 2013; Gehrke et al., 2011) alternatively, sudden spikes could indicate a stress response. While the heart rate monitors collected data every two seconds, the software exported the average and maximum heart rate.

Statistical Analysis

All statistical analyses were performed using RStudio, version 2024.04.1+748.

GREAT Scale

Each corresponding set of two scores given by the RA was averaged. The baseline scores for students with missing data (i.e., absent on that date) were imputed by using the first available score. Similarly, final session scores were assigned, generally by using the scores from the final week of data collection, but were imputed with the final available scores when necessary. In addition to the first and final session scores, the differences between the scores were also computed. Data was partitioned into two groups: the adaptive riding only (AR) control group and the adaptive riding with music therapy (MT/AR) intervention group.

Descriptive statistics were found for each group. Normality for each group was assessed via visual analysis, Q-Q plots, and the Shapiro-Wilk test. Analogous AR and MT/AR groups were compared using a two-sample t-test if both groups were normal (Welch if non-equal variance, pooled if equal variance) or a Mann-Whitney test if at least one group was not normal (Table 1). One-sided tests were used under the assumption that average scores given by RAs in the MT/AR group were lower than scores given by RAs in the AR group. The exception was when comparing the first session data between the two groups using a two-sided test, since it was not hypothesized that a difference would be detected in the first session.

Table 1

Descriptive Statistics of GREAT Score Sums

Groups	Min	Q1	Med	Mean (SD)	Q3	Max
Combined						
Baseline, SOS	1.50	3.75	5.50	5.37 (2.43)	6.50	10.00
Baseline, EOS	0.50	3.00	5.00	4.80 (2.56)	6.00	10.00
Final score, SOS	0.00	1.25	2.00	2.57 (2.21)	3.25	8.00
Final score, EOS	0.00	0.25	0.50	1.23 (1.28)	2.00	3.50
MT						
Baseline, SOS	2.00	4.00	6.00	5.50 (2.22)	6.50	9.00
Baseline, EOS	3.00	5.00	6.00	5.78 (2.22)	6.00	10.00
Final score, SOS	0.00	1.50	1.50	1.89 (1.24)	3.00	3.50
Final score, EOS	0.00	0.00	0.50	1.00 (1.09)	2.00	3.00
AR						
Baseline, SOS	1.50	3.63	4.75	5.17 (2.93)	6.25	10.00
Baseline, EOS	0.50	1.75	3.25	3.33 (2.46)	4.00	7.50
Final score, SOS	0.50	1.38	2.75	3.58 (3.02)	5.625	8.00
Final score, EOS	0.00	0.50	1.00	1.58 (1.56)	3.00	3.50
Differences						
MT/AR, SOS	-6.00	-5.50	-5.00	-3.61 (2.42)	-1.50	-0.50
MT/AR, EOS	-8.00	-6.00	-4.50	-4.78 (2.09)	-4.00	-1.00
AR, SOS	-5.00	-3.38	-1.75	-1.58 (2.64)	0.63	1.50
AR, EOS	-4.00	-3.25	-2.00	-1.75 (1.97)	-0.38	1.00

SOS = Start of Session; EOS = End of Session; MT = music therapy (intervention); AR Adaptive Riding (control)

Global Assessment Scale

The analogous scores given by the RAs were averaged. Separate analyses were performed on the average GAS profile score given by the RAs and CTRIs. Each participant was assigned first session profile scores and final session profile scores. Typically, these were the rider's profile score from the first week and last week of data collection, respectively. In cases where students were absent for the first week of their session, the first session profile scores were imputed by using their first available score. Similarly, missing final session scores were imputed by using the final available scores, when necessary. Multivariate regression and multiple linear regression models were run, treating final goal scores as the outcome variables and initial goal scores, treatment group, and number of sessions in which a student participated as input variables. Demographic information was built into the models as covariates. Separate models were built from data gathered by RAs and those gathered by instructors.

Heart Rate

Heart rate data were collected only for the duration of the riding session. Data were analyzed for differences between treatment groups and for trends more broadly. Descriptive statistics were found, and groups were assessed for normality via Q-Q plots and the Shapiro-Wilk test. Analogous AR and AR/MT groups were compared using a two-sample t-test if both groups were normal (Welch if non-equal variance, pooled if equal variance) or a Mann-Whitney test if at least one group was not normal. Data for instructors (n=2) and Music Therapist (n=1) were not analyzed.

Results

Demographics

Students were randomly assigned to the control (adaptive riding, or AR; n = 7) or intervention group (adaptive riding with music therapy, or MT/AR; n = 9). Once established, the division of groups was consistent through all eight weeks.

Of the sixteen students, fifteen were African American and one was Caucasian. Ten were male and six were female. Two were diagnosed with Attention-Deficit/Hyperactivity Disorder, two with Autism Spectrum Disorder, and six with behavioral health disabilities. All were between the ages of ten and twelve and came from classrooms within the same public school.

Of the 40-horse herd, 27 were involved in this study. Four of the miniature horses were led by students who were not eligible or who chose not to ride. All other horses were ridden in English saddles or bareback pads as assigned by the CTRI. All horses wore a bitless bridle with the student's reins attached to the side rings, and the lead rope attached underneath. Two horses also had a bit, though no reins were attached to the bit directly.

GREAT Scale

Both AR and MT/AR groups saw improvements in GREAT scores. The downward (positive) trend in scores over time can be seen in the MT/AR EOS boxplots (Figure 1). A difference score was obtained by subtracting the baseline from the final. Descriptive statistics of these variables are shown in Figure 1.

The MT/AR group improved an average of five points on their End of Session (EOS) GREAT scores, whereas the AR group saw only an average improvement of 1.75 points on the analogous measurement. When comparing GREAT scores, the MT/AR group's EOS demonstrated statistically significant improvement, indicated by higher scores, relative to the AR (control) group.

Global Assessment Scale

The regression models yielded no statistically significant findings between AR and MT/AR groups. The scores were averaged, and separate analyses were performed on the RA-assigned GAS compared to the instructor-assigned GAS. RA-assigned scores over time generally trended upward (positive trend, Figure 2), as opposed to the instructors' scores, which remained more neutral over time.

Figure 1
Boxplots of GREAT Scale Scores Over Time

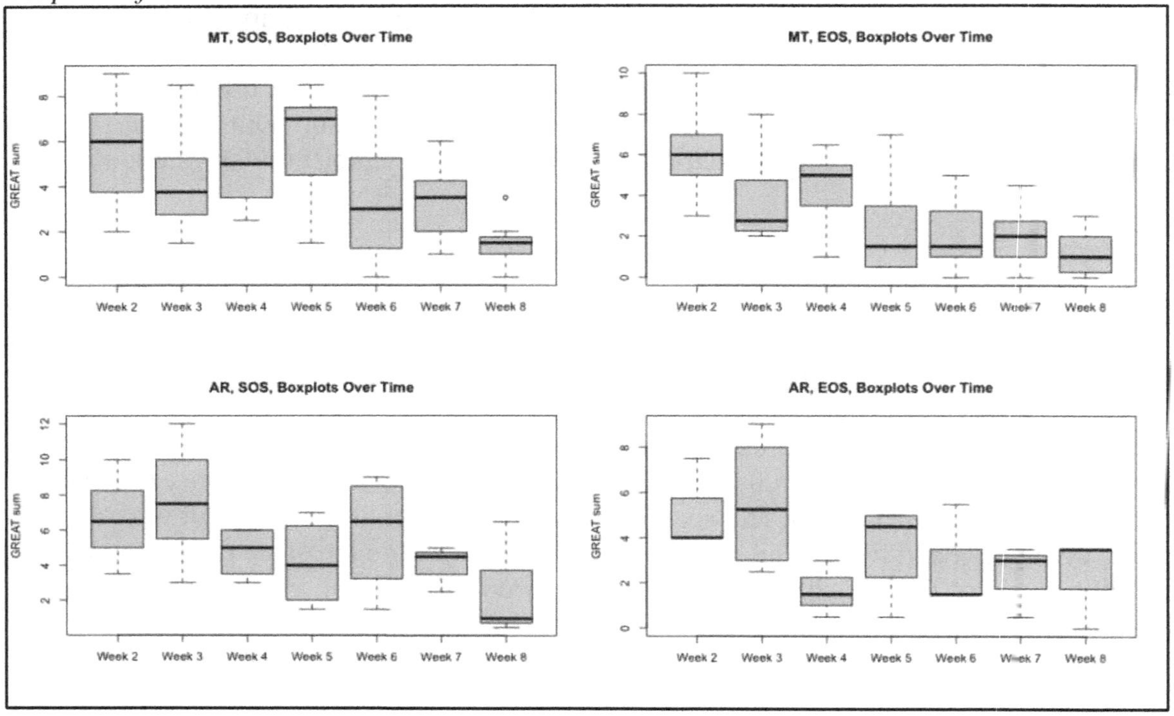

Heart Rate

Heart rate data were analyzed for differences between treatment groups and for trends more broadly. Descriptive statistics are given in Table 2. Boxplots of treatment groups are shown in Figure 3.

Table 2
Statistical Analysis of GREAT Sum Scores between MT/AR vs AR Groups

Group	Test (Direction)	Test Statistic (p-Value)
First session, SOS	t-Test, Pooled (Two-Sided)	t = 0.237 (0.818)
First session, EOS	t-Test, Pooled (Two-Sided)	t = 1.957 (0.079)
Final session, SOS	Welch's t-Test (Left)	t = -1.301 (0.120)
Final Session, EOS	t-Test, Pooled (Left)	t = -0.795 (0.225)
Differences, SOS	Mann-Whitney (Left)	W = 13.5 (0.062)
Differences, EOS	t-Test, Pooled (Left)	t = -2.945 (0.008)*

*p-value significant at the 0.05 level
SOS = Start of Session; EOS = End of Session

Figure 2
Boxplots of GAS Scores Over Time for RAs

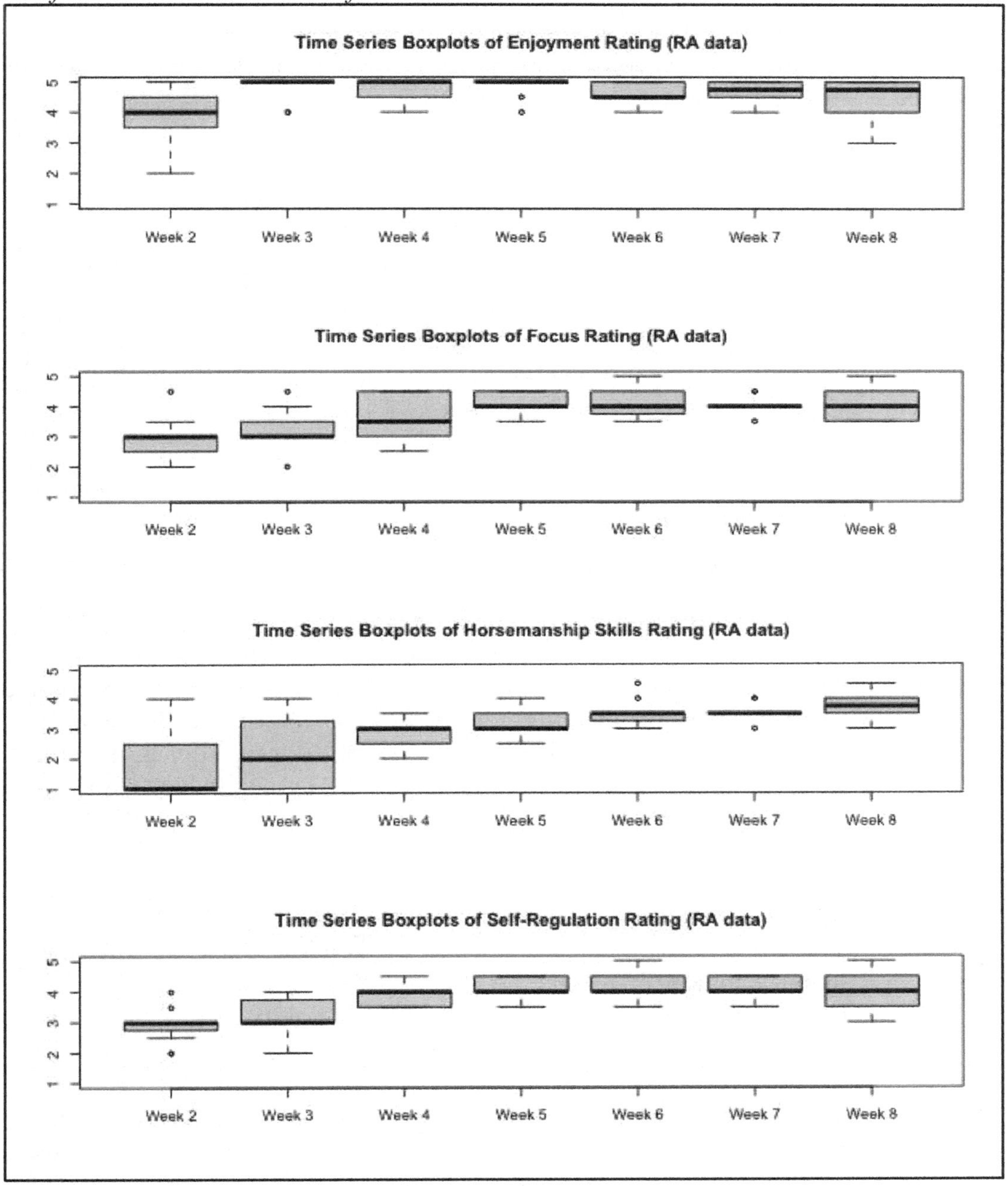

Figure 3

Boxplots comparing average heart rates of music therapy (intervention) and adaptive riding (control) groups.

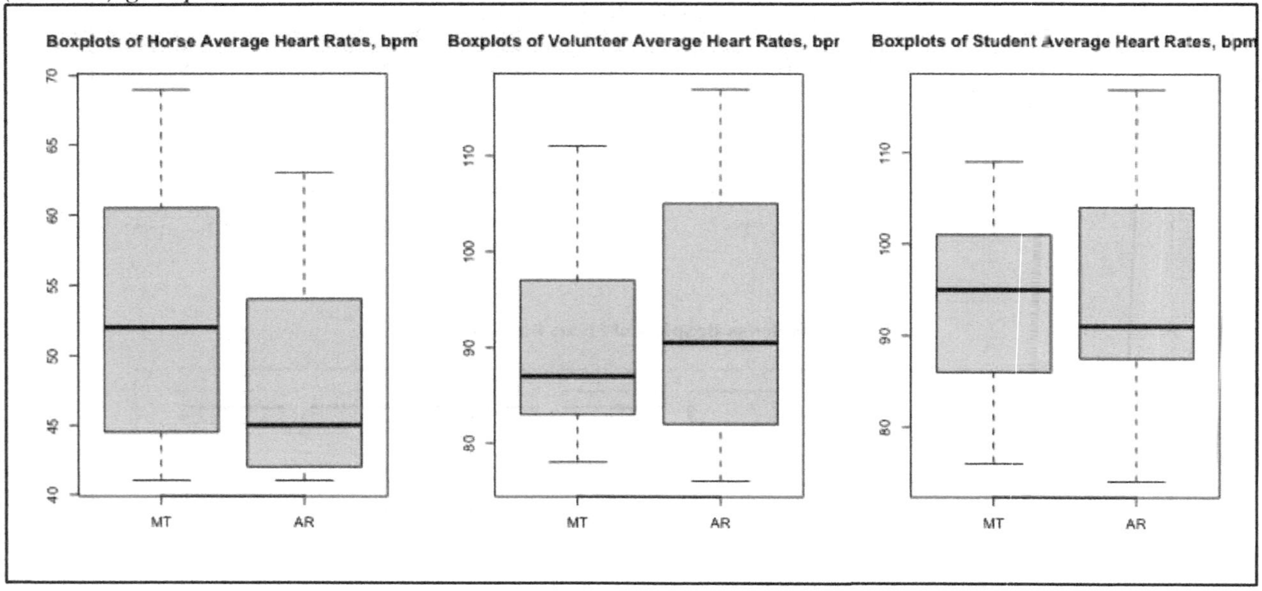

No statistically significant differences were found in the average heart rates of the groups (Table 3). Data also explored changes by session (i.e., week), but no trends across time were found. There were trending data that could have been indicative of a difference in the horse's heart rate, but the sample size was too small for appropriate tests to distinguish statistical significance. If the horse's heart rate difference had been statistically significant, it would have been lower in the control (AR), than in the intervention (MT/AR) group, which is the opposite of the hypothesis.

Table 3

Statistical Analysis of Difference in heart rate between MT/AR and AR groups

	Test Used	Test Statistic	p-value
Horse	Mann-Whitney, two-sided	W = 228.5	0.147
Students	Pooled t-test, left	t = -0.476	0.318
Volunteers	Mann-Whitney, left	W = 172	0.321

The distributions of average heart rate were right-skewed (higher rate) for horses and volunteers, but demonstrated more normality for the students. Figure 4 depicts the trends of the heart rate, but does not distinguish between treatment groups because there was no difference between groups. The heart rate trend evident in the horses was not a statistically significant finding. However, this may have been because the number of horse was insufficient, and thus the sample size was potentially statistically underpowered.

Figure 4

Average heart rate by participant group

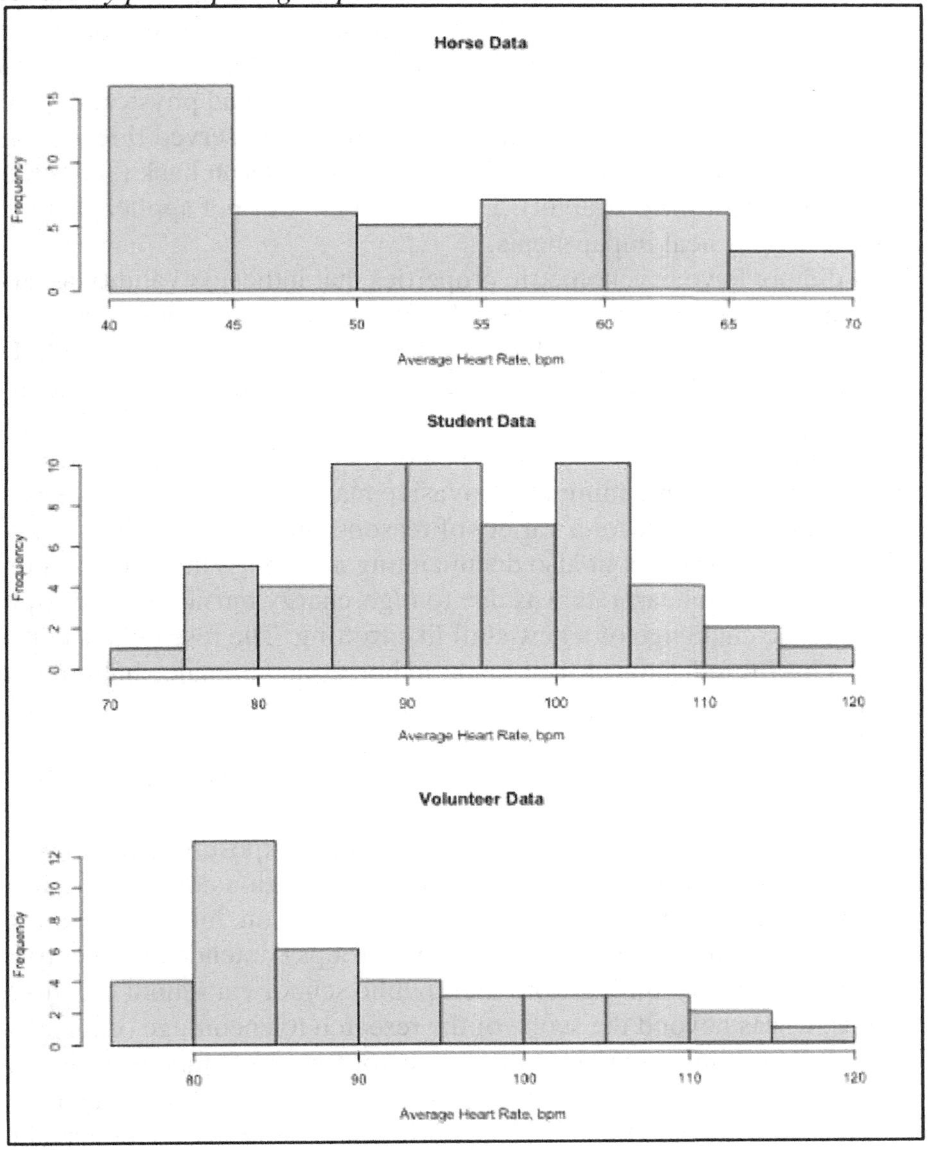

Discussion

The findings demonstrate that the addition of music therapy to adaptive riding lessons may offer increased benefit to students engaged in adaptive riding in areas of focus, horsemanship, self-regulation, and enjoyment. To explore the potential impact and implications of the findings, analysis should be considered in two ways: 1. Effectiveness of the measures for addressing the research questions; 2. Effectiveness of the intervention for addressing the research questions.

Effectiveness of the Measures

At the outset of this study, researchers were aware that a discussion of the measures would be pertinent, as the field of EAS has had a dearth of measures. Often, researchers have used measures developed from other fields simply because EAS-specific measures did not exist. It was surprising that the GREAT Scale was not used in more studies as a postural scale for horseback riding, as this scale was effective in identifying change over time.

The GREAT scale was limited in inclusivity as the image was of an able-bodied, thin white silhouette. The researchers were challenged to determine postural alignment when the rider was larger or shaped differently than the silhouette, or when the rider had physical disabilities that were not represented by the scale. Additionally, the researchers observed riders in positions not indicated on the scale. For example, some participants tended to lean back (shoulders far behind their hips, while keeping their back straight). Thus, this scale was not applicable to riders who have musculoskeletal or physical impairments.

While the GAS did not have psychometric properties that indicated validity as a complete sub-scale, this was not necessarily an indication that the scale should not be used. The inclusion of the scale was an effort to explore the validity of the measure for use in research. Using the measure, the CTRIs and RAs were able to reliably track student change over time, relative to the individual student and group goals.

Heart rate was included as a physiological indicator that could be captured in a cost-effective manner, and more importantly, in a minimally invasive manner. The challenge was that heart rate could vary moment to moment for a variety of reasons, and therefore, the data was not descriptive for horse or human without also documenting and analyzing context. For example, it was unknown if an increase in heart rate was due to high-energy music, a horse frightened by a sudden movement, or the challenge of a new skill like trotting. The researchers concluded that heart rate could be a useful indicator of well-being in horses and humans, but not without additional context.

Effectiveness of the Intervention

Observations of the groups were over two eight-week sessions. The AR and MT/AR lessons were separated by a thick fabric divider. Only the MT/AR group was intended to hear and engage with the music. The AR group may have had minimal exposure to the music. However, it was confirmed that the lyrics, instruction, or associated information could not be heard. Researchers acknowledge the fabric wall was a possible limitation, but not a limiting variable.

As indicated throughout this study, students in both groups contended with chronic absenteeism. The students were coming from their public school via school organized transportation. Thus, it was beyond the scope of the research to encourage or mandate attendance in the study. Chronic absenteeism may have been a factor for some students based on societal issues such as poverty. The impact of missing school lead to disrupted learning and complications for the study of the efficacy of supportive interventions, as indicated herein. The U.S. Department of Education (2023) noted that two-thirds of K-12 students, or one in five, missed almost four weeks throughout the academic year. Chronic absences have caused negative long-term impacts on the students' academic achievement, mental and physical health, social engagement, and intervention engagement. Evidence supports that EAS has the potential to positively impact the development of self-esteem, confidence, and feelings of life satisfaction for participating students (Carlsson, et al., 2024). However, students who were absent from school are also absent from supportive interventions.

While additional data was collected on the horses, it was not salient to the research questions and findings, thus not included herein. Data that was collected but not relevant to the research questions included: herd size and frequency of work, lesson duration, and fit and use of tack, and instructor feedback about horses in session(s).

Rhythmic Entrainment was an effective theoretical framework for this pilot study. As postulated, the theory may positively impact psychosomatic experiences, including heartbeat, among participants in music therapy (Bharathi et al., 2019; Bruscia, 1988). Moreover, Reybrouch and colleagues (2021) postulated that music may serve as a tool for connection, activation and regulating (Bonassi et al., 2023). By stimulating psychosocial experiences, heart rates may be affected. Though based on the data collected, future research will be needed to extrapolate why and by what means.

Behavior Theory was appropriately applied to the integration of music therapy in adaptive riding through the use of positive reinforcement for learning the skills, trying new and challenging tasks, and incentivizing engagement (Bandura et al., 1977). Consistent with the findings of Bunketorp-Käll and colleagues (2019), adaptive riding demonstrated positive outcomes in horsemanship skills, focus, enjoyment, and self-regulation. However, as noted by Gonzalez-Hoelling and colleagues (2024), "care should be taken in interpreting and generalizing findings" (p. 1).

Limitations

This pilot study sought to use established measures as available, and integration of appropriate technology (e.g., polar sensors) without posing an imposition to the CTRIs, students, horses, and Center. All measures used were new to the research Center, CTRIs, and Ras, thus there was a substantial learning curve to test and teach the measures. The challenges of on-board multiple new measures further highlighted the need for standardized measures crafted by and for the EAS industry. Through the process of data analysis, limitations of current data and opportunities for future research emerged, including hertz rate of songs, variability of heart rate, and other indicators to consider for potential significance.

The researchers found heart rate a challenge to use as an independent measure because there were too many environmental factors to parse out meaning. Without the context of what was happening during the session, heart rate was an isolated measure that was not reliable. The heart rate trend evident in the horses was not a statistically significant finding, but was potentially underpowered and could be explored in further research. The ability to use heart rate as a physiological indicator, positive or negative, was possible because the monitors were more cost-effective and less invasive than many other physiological indicators (i.e., saliva, blood, or urine). Even so, heart rate monitor cost may be prohibitive in minimally or non-funded research.

Consistent with the majority of EAS research, a larger sample size of students, volunteers, instructors, and horses would have been ideal for meaningful data analysis that could detect statistical significance. Most evident in this pilot study was the sample size of the horses for measuring heart rate, because there was potential for statistically significant findings had the data not been underpowered.

Adaptive riding lessons differ in dosage, duration, population served, and are inconsistent regarding controls (i.e., volunteers, horses, lesson content). As such, researchers must establish clear controls, ideally by using standardized measures, and share transparently about null findings. Doing so is the only way to move research forward and advance collective knowledge.

Future Research

Considering that the measures in this pilot study included physiological data, it may have been advantageous to consider physiological theories such as polyvagal theory or the biophilia hypothesis to better understand the underlying mechanisms of change that affected intervention efficacy.

While the researchers demonstrated competence and efficiency, there was still concern about human error from trying to observe multiple variables simultaneously. Future research could remedy this by video recording the sessions, then watching and rating recordings at a later time. Similarly, this pilot study demonstrated the fidelity of using heart rate monitors and the collection of minimal amounts of data. Future research could delve deeper into raw data and data analysis (e.g., variability). Also recommended could be to include participants in the data collection through mini-interview, short-answer self-report, or pre/post focus groups. Collection of such qualitative data could be beneficial for students with disabilities as explicit inclusion could reinforce their sense of autonomy.

In the absence of a standardized scale for EAS progress tracking, the CTRI could use the GAS as a tool to track their students' progress from session to session. Since the scale and subscales do not hang together statistically, the CTRI could have changed the language of the goals and adapted the scale as desired. As such, EAS Centers may find the GAS to be a unified way of tracking student progress. Reflecting on Smith Hybels (2015) presentation, this clinical use is aligned with the original intention of the measure. Similarly, another opportunity is to consider adaptation of the GREAT scale for more riding postures (e.g., different disciplines) and diverse body types, abilities, and sizes.

For future research it may be advantageous to explore the effect of music therapy in non-riding interventions (i.e., equine-assisted learning and psychotherapy). Finally, dedication to the development of measures that are culturally relevant and affirming of all body types, with strong psychometric properties, would be a significant contribution to the EAS field.

Conclusion

All participants experienced improvements in posture as measured by the GREAT scale, however improvements were found to be significantly higher for the intervention group. Improvements in focus, horsemanship/posture, self-regulation, and enjoyment, measured by the GAS scale, were seen in both groups, but with no statistically significant difference. The GAS was not effective in identifying statistically significant change over time, however it did merit clinical utility that should be considered by riding instructors. The addition of music therapy to adaptive riding lessons may offer increased benefit to students engaged in adaptive riding in areas of focus, horsemanship, self-regulation, and enjoyment.

Acknowledgments: We acknowledge the RAs who supported the research (Alexis Koran and Abigail McNaughton), CTRIs (Lauren Simak and Tonya Zimmer), Music Therapist (Christine O'Keefe), the volunteers who give of their time every week, and the horses.

Data Availability Statement: Data available on request from the corresponding author.

Conflict of Interest: The authors declare no conflicts of interest. The funders had no role in the design of the study; the collection, analyses, or interpretation of data; in the writing of the manuscript, or in the decision to publish the results.

Funds: This research was funded through a gift to Fieldstone Farm Therapeutic Riding Center.

Author Contributions: Conceptualization, AV; methodology, AV; formal analysis, AV, RR, DH, LQ,; investigation, AV, EE, AWB, CL; resources, AV; data curation, AV, EE, AWB, CL;

writing-original draft preparation, AV, EE, AWB, CL, DH, RR, LQ.; writing-review and editing, AV, EE, LQ; project administration AV. All authors have read and agreed to the published version of the manuscript.

Statement: This study was conducted in accordance with the guidelines of the Declaration of Helsinki, and approved by the Institutional Review Board of Cleveland State University (IRB-FY2024-24)

References

American Music Therapy Association. (n.d.). About music therapy and AMTA. American Music Therapy Association. https://www.musictherapy.org/about/

Bandura, A., Adams, N. E., & Beyer, J. (1977). Cognitive processes mediating behavioral change. *Journal of Personality and Social Psychology, 35*(3), 125–139. https://doi.org/10.1037/0022-3514.35.3.125

Bharathi, G., Jayaramayya, K., Balasubramanian, V., & Vellingiri, B. (2019). The potential role of rhythmic entrainment and music therapy intervention for individuals with autism spectrum disorders. *Journal of Exercise Rehabilitation, 15*(2), 180–186. https://doi.org/10.12965/jer.1836578.289

Bitschnaul, C., Jones, J., Haldi, J., Laukkanen, R., & Weishaupt, M. (2014). *Polar sport zones for horses* [White paper]. https://www.polar.com/sites/default/files/static/science/white-papers/polar-sport-zones-for-horses-whitepaper.pdf

Bonassi, G., Lagravinese, G., Bove, M., Bisio, A., Botta, A., Putzolu, M., Cosentino, C., Mezzarobba, S., Pelosin, E., & Avanzino, L. (2023). How Music Moves Us: Music-induced Emotion Influences Motor Learning. *Neuroscience*, *526*, 246–255. https://doi.org/10.1016/j.neuroscience.2023.06.023

Bruscia, K. E. (1988). A survey of treatment procedures in improvisational music therapy. *Psychology of Music, 16*(1), 10–24. https://doi.org/10.1177/0305735688161002

Bunketorp-Käll, L., Pekna, M., Pekny, M., Blomstrand, C., & Nilsson, M. (2019). Effects of horse-riding therapy and rhythm and music-based therapy on functional mobility in late phase after stroke. *NeuroRehabilitation*, *45*(4), 483–492. https://doi.org/10.3233/NRE-192905

Carlsson, I. M., Bräutigam Ewe, M., Nymberg, P., & Jormfeldt, H. (2024). Building up bit by bit, parents' experiences of equine-assisted intervention among children and adolescents with mental illness: a grounded theory study. *International journal of qualitative studies on health and well-being*, *19*(1), 2354945. https://doi.org/10.1080/17482631.2024.2354945

Ferlazzo, A., Fazio, E., & Medica, P. (2024). The relevance of auditory perception and the effects of music on horses. *Atti della Accademia Peloritana dei Pericolanti-Classe di Scienze Medico-Biologiche*, *112*(1), 1-10. https://doi.org/10.13129/1828-6550/APMB.112.1.2024.SD1

Frease, L. (1996) The G.R.E.A.T. Postural Scale. University of Florida.

Gehrke, E. K., Baldwin, A., & Schiltz, P. M. (2011). Heart rate variability in horses engaged in equine-assisted activities. *Journal of Equine Veterinary Science, 31*(2), 78-84. https://doi.org/10.1016/j.jevs.2010.12.007

Gonzalez-Hoelling, S., Reig-García, G., Bertran-Noguer, C., & Suñer-Soler, R. (2024). The effects of rhythmic auditory stimulation on functional ambulation after stroke: a systematic review. *BMC complementary medicine and therapies*, *24*(1), 45. https://doi.org/10.1186/s12906-023-04310-3

Gueguen, L., Henry, S., Delbos, M., Lemasson, A., & Hausberger, M. (2023). Selected acoustic frequencies have a positive impact on behavioural and physiological welfare indicators in thoroughbred racehorses. *Animals*, *13*(18), 2970. https://doi.org/10.3390/ani13182970

Hallberg, L. (2018). *The clinical practice of equine-assisted therapy: Including horses in human healthcare.* Routledge. IBSN: 978-1138674639

Jabborova, D. & Mohirabonu, Y. (2023). Effects of music on human health. *Social Science and Innovation*, *1*(3), 6-10. https://bestjournalup.com/index.php/ssai/article/view/152. ISSN 2992-9180.

Kaplan, R. S., & Steele, A. L. (2005). An analysis of music therapy program goals and outcomes for clients with diagnoses on the autism spectrum. Journal of music therapy, 42(1), 2–19. https://doi.org/10.1093/jmt/42.1.2 Kapteijn, C. M., Frippiat, T., van Beckhoven, C., van Lith, H. A., Endenburg, N., Vermetten, E., & Rodenburg, T. B. (2022). Measuring heart rate variability using a heart rate monitor in horses (*Equus caballus*) during groundwork. *Frontiers in Veterinary Science*, *9*, 939534. https://doi.org/10.3389/fvets.2022.939534

Merrigan, J. J., Stovall, J. H., Stone, J. D., Stephenson, M., Finomore, V. S., & Hagen, J. A. (2023. Validation of Garmin and Polar Devices for Continuous Heart Rate Monitoring During Common Training Movements in Tactical Populations. *Measurement in Physical Education and Exercise Science*, *27*(3), 234–247. https://doi.org/10.1080/1091367X.2022.2161820

Navalta, J. W., Davis, D. W., Malek, E. M., Carrier, B., Bodell, N. G., Manning, J. W., Cowley, J., Funk, M., Lawrence, M. M., & DeBeliso, M. (2023). Heart rate processing algorithms and exercise duration on reliability and validity decisions in biceps-worn Polar Verity Sense and OH1 wearables. *Scientific reports*, *13*(1), 11736. https://doi.org/10.1038/s41598-023-38329-w

Paré, G., & Kitsiou, S. (2017). Methods for literature reviews. In F. Lau & C. Kuziemsky (Eds.), Handbook of eHealth evaluation: An evidence-based approach. University of Victoria. https://www.ncbi.nlm.nih.gov/books/NBK481583/

PATH Intl. (2025). About PATH Intl. https://pathintl.org/about/

Polar (2022. Polar Optical Heart Rate [White paper]. https://www.polar.com/en/img/static/whitepapers/pdf/ -optical-heart-rate-white-paper.pdf?srsltid=AfmBOoobYAHd_XOKth0bVofQE-VEjmyGtD2ZbLUSUx-9e0EltJrL5oiT

Pruvost, M., Bellone, R., Benecke, N., Sandoval-Castellanos, E., Cieslak, M., Kuznetsova, T., ... & Ludwig, A. (2011). Genotypes of predomestic horses match phenotypes painted in Paleolithic works of cave art. Proceedings of the National academy of sciences, 108(46), 18626-18630. https://doi.org/10.1073/pnas.1108982108

Reybrouck, M., Podlipniak, P., & Welch, D. (2021). Music listening and homeostatic regulation: Surviving and flourishing in a sonic world. *International journal of environmental research and public health*, *19*(1), 278. https://doi.org/10.3390/ijerph19010278

Rossi, S. L. (2018). *Interdisciplinary collaboration between music therapy and therapeutic riding: Feasibility and design* [Doctoral dissertation, Sam Houston State University]. Texas Digital Library. https://shsu-ir.tdl.org/server/api/core/bitstreams/92f39eeb-95e4-4487-9bee-ebda324101d1/content

Schulkin, J., & Raglan, G. B. (2014). The evolution of music and human social capability. *Frontiers in neuroscience*, *8*, 292. https://doi.org/10.3389/fnins.2014.00292.Smith Hybels, C. (2015) Five EAS facilities piloted the Global Assessment Scale. Power-point Presentation. Conference/presentation: unknown.

U.S Department of Education. (2023, October 12). *Chronic absenteeism challenges schools.* John Hopkins University Hub. https://hub.jhu.edu/2023/10/12/chronic-absenteeism-challenges-schools/

Thompson, F., Ketcham, C. J., & Hall, E. E. (2014). Hippotherapy in children with developmental delays: Physical function and psychological benefits. *Advances in Physical Education, 2014*. http://dx.doi.org/10.4236/ape.2014.42009

von Lewinski, M., Biau, S., Erber, R., Ille, N., Aurich, J., Faure, J. M., ... & Aurich, C. (2013). Cortisol release, heart rate and heart rate variability in the horse and its rider: Different responses to training and performance. The Veterinary Journal, 197(2), 229-232. https://doi.org/10.1016/j.tvjl.2012.12.025

Wood, W., Alm, K., Benjamin, J., Thomas, L., Anderson, D., Pohl, L., & Kane, M. (2021). Optimal terminology for services in the United States that incorporate horses to benefit people: A consensus document. *The Journal of Alternative and Complementary Medicine*, *27*(1), 88-95. https://doi.org/10.1089/acm.2020.0415

A multiple case-study analysis of Autistic experiences with therapeutic horseback riding

Tiffany Banks, Assist. Prof., Nancy Kusmaul, Prof., and Jossalin Hughes, BSW, Student

School of Social Work, University of Maryland Baltimore County

All correspondence and requests for reprints should be addressed to:
Tiffany Banks at
Tbanks4@umbc.edu or
University of Maryland, Baltimore County
Sherman Hall-B, 327
1000 Hilltop Circle, Baltimore, USA

Handling Editor: Alexandra Stergiou
Reviewers: Ann Hemingway & Rebecca Stares

Volume 25, Number 1, January 2026

DOI: https://doi.org/10.58862/JORO2362

The use of animal-assisted interventions (AAI) to improve the lives of Autistic individuals is a growing field, but still a treatment option that is considered complementary to evidence-based practices. One form of AAI, therapeutic horseback riding (THR), has shown promise for the Autistic community. Benefits of THR include increased social communication, decreased aberrant behaviors, and increased mood regulation, but why and how these programs work remains understudied. In this multiple-case study comparative analysis, we explore the lived experiences of three Autistic youth engaged in a THR program, focusing on how their experiences shape their lives. Using an ecological systems approach to organizing data, three themes emerge: voice, individualization, and the personal experience of difference. Through interviews, the youth shared distinct experiences with self-advocacy, individual accommodations, and feelings of marginalization. Their stories suggest that accommodation and positive regard, even in universally designed settings, play an important role in the experience of THR programs. Implications and limitations are discussed.

Keywords: Autism, Lived Experience, Therapeutic Horseback Riding

Autism Spectrum Disorder (ASD) is a neurocognitive developmental disability that affects the use of social communication skills and is marked by the use of restricted and repetitive behaviors (American Psychiatric Association [APA], 2013). While Autistic individuals make up around 2% of the general population (Maenner et al., 2023), the mental health community reports that Autistic clients make up a much larger percentage of their population (Belardinelli et al., 2016; Buck et al., 2014; Rosen et al., 2018). The prevalence of co-occurring psychiatric conditions with ASD highlights an important intersection for practitioners, diagnosticians, and researchers throughout the mental health field.

One area of mental health practice that has shown promising outcomes for Autistic individuals is the field of animal-assisted interventions (AAI). AAI is the use of various animals to support goal directed and non-goal directed activities to benefit the lives of humans (American Veterinary Medical Association, 2025). Literature reviews on the subject indicate positive correlations between social and emotional outcomes for people with Autism and the incorporation of animals into the helping process (Davis et al., 2015; Hoagwood et al., 2017; Nieforth et al., 2021), with some of the strongest evidence supporting a specific model called equine assisted intervention.

O'Haire (2013) wrote that a complication in the field of AAI is the inconsistent use of language throughout the literature. For clarity, this paper uses the term equine assisted services (EAS) as a term in which AAI is being conducted with the support of a horse (Professional Association of Therapeutic Horsemanship [PATH] International, 2023). Within EAS, there are three terms to further differentiate the experiences: therapy, learning, and horsemanship. This paper explores the impact of a therapeutic horseback riding (THR) program. THR is non-goal oriented and typically led by a riding instructor, focusing on riding skills with therapeutic gains being a secondary outcome, thus falling into the horsemanship category of EAS (PATH International, 2023).

Studies demonstrate that THR may decrease hyperactivity and irritability and increase social abilities and language use (Gabriels et al., 2015, 2018), improve social communication skills (Coman et al., 2018; Ward et al., 2013), increase quality of life (Lovrić et al., 2020), decrease stress (Kemeny et al., 2022), and reduce behavioral concerns (Holm et al., 2014; Zoccante et al., 2021). While these studies have limitations, there is promise as the field increases methodological rigor to support the efficacy of AAIs (Nieforth et al., 2021).

While studies are beginning to incorporate physiological measurements into AAI research to objectively demonstrate internal participants' response patterns (Rodriguez et al., 2018), some early studies have been inconclusive (Pan et al., 2019; Protopopova et al., 2019). This is further complicated by the variability in the release of hormones in Autistic children (Corbett et al., 2009, 2010, 2016), creating reliability concerns for measure. Furthermore, the use of physiological measurements should not replace input from Autistic participants regarding their lived experience, which is naturally subjective. Subsequently, researchers still cannot confirm why or how AAI, especially non-therapeutic approaches like THR, creates the outcomes associated with participation in these programs.

According to Dr. Temple Grandin, an Autistic self-advocate and renowned equine expert, the benefits associated with AAI are potentially created by the shared communication patterns between Autistic humans and animals (Grandin et al., 2015; Grandin & Johnson, 2006). Using her own lived experiences and expertise, Grandin hypothesizes that animals and Autistic people both have "sensory-based thinking", which may provide relief from the complex demands of socializing. One way to better understand how AAI and THR are impacting individuals is to use diverse methodological approaches to knowledge to triangulate data from multiple sources (Denzin, 2010) and to explore case studies of other Autistic individuals to see if their experiences align with Grandin's proposals.

Current Literature

The current project was planned to be embedded within an established manualized program (Merritt & Gabriels, 2020), thus fully understanding the efficacy and outcomes associated with this established model are essential. Publications on previous research projects that utilized this model indicate promising outcomes in the areas of aberrant behaviors, speech and language, as well as social and emotional wellness (Gabriels et al., 2015, 2018; Pan et al., 2019). Gabriels et al. (2012) conducted their initial pilot study of the previously mentioned 10-week THR intervention with Autistic participants (n = 42) compared to Autistic participants (n = 16) in waitlist comparison. The Gabriels et al. (2012) study found correlations between THR and a reduction in irritability, lethargy, stereotypic behaviors, and hyperactivity, while increasing observations of expressive language skills.

The follow-up RCT conducted by Gabriels et al., (2015) involved randomly assigning Autistic participants (n = 116) ages 6 to 16 years to one of two groups: therapeutic horseback riding (THR) or no-horse contact in the Barn Activity (BA) control. Caregiver ratings on pre- and post- intervention assessments showed significant improvements in participants involved in the THR group compared to the BA control on a caregiver report measure of irritability and hyperactivity behaviors. Speech and language pathologists, blind to the intervention assignment of participants, completed pre- and post-intervention standardized language sample using the Systematic Analysis of Language Transcripts (SALT). Participants in the THR group showed significant improvements in word usage on the language sample compared to the BA control participants. Participants (n = 64) from the Gabriels et al., (2015) study participated in a six-month follow-up and showed sustained improvements in irritability and word usage (Gabriels et al., 2018).

The Gabriels research team then replicated the ten-week program at a different riding center to compare outcomes (Pan et al., 2019). This smaller study included a sample of Autistic youth (n = 16) randomized to one of two groups: THR or BA control. This replication study added another measurement to the battery of tests, salivary cortisol. Outcomes of this pilot study were similar in terms of the THR participants making significant improvements in caregiver-reported irritability and hyperactivity behaviors compared to the BA control, but the THR group did not replicate other previous findings from Gabriels et al., (2015). The use of salivary cortisol to measure physiological changes produced minimal results in this pilot study.

Across these studies, a reduction in irritability appears to be the most consistent outcome, with sustainable gains. The active RCT of THR for Autistic youth with co-occurring mental health symptoms increases methodological rigor through the integration of physiological measures, including salivary cortisol, a heart rate monitor, and electrodermal activity monitoring. Additionally, the study from the Gabriels team has two sites in different parts of the United States and a goal of recruiting 250 participants. The longevity and growth of evidence supporting the effectiveness of this 10-week manualized program suggest great promise for the Autistic population. This paper presents additional context to the larger RCT by adding the Autistic participant voice and lived experience to a growing body of evidence.

This qualitative study, designed by the first author and embedded into the RCT being conducted by the Gabriels team, seeks to understand the first-hand experiences of Autistic youth participating in THR by inviting their feedback, which has been historically absent from the investigative measures used to analyze outcomes in previously published literature.

Method

To triangulate lived experience with objective data, this qualitative project utilized a feminist framework to explore the lived experiences of three Autistic youth engaged in a larger RCT of a THR program (1R01HD097693-01A1; ClinicalTrials.Gov NCT #: NCT04606966). By embedding a qualitative project inside of a preexisting program with a decade of evidence to support its outcomes, we can better consider the research question at hand: What are the experiences of Autistic youth who engage in THR, and how do those experiences impact their lives?

Participants

Participants in this qualitative study were members of a single cohort at a single site of the larger RCT study. As part of the larger study's procedures, participants' ASD diagnoses were confirmed using the Social Communication Questionnaire (SCQ; Bölte et al., 2008) and the Autism Diagnostic Observation Schedule 2 (ADOS-2; Lord et al., 2012). Participants were included in this qualitative study if they had verbal fluency as measured by the third module of the ADOS-2. The targeted cohort consisted of six participants, three of whom were randomly assigned to the treatment group and three of whom were randomly assigned to the control group. Of these six individuals, four met the verbal requirements for oral interviewing. These four participants and their caregivers were provided an Institutional Review Board (IRB) approved recruitment flyer. The primary author met with the four youth and their caregivers at the riding center, during the collection of baseline assessment data. Three of the participants agreed to participate and completed an informed consent/assent process. Guardians and youth over the age of thirteen reviewed a consent form with the primary author, explaining the study procedure and associated risks. Youth under the age of thirteen completed an assent process that utilized developmentally appropriate language, only after their guardian consented to their participation. The three participants were a six-year-old male and twelve-year-old male from the THR group and a sixteen-year-old female in the BA group.

Procedure

This study was reviewed and approved by the Institutional Review Board (IRB) at Colorado State University (#3500) and University of Maryland Baltimore County (#1399). Participants in this qualitative study engaged in the larger RCT study's 10-week THR group or BA control group, both of which consisted of an equine learning activities curriculum (Burnside & Gabriels, 2019; Merritt & Gabriels, 2020). The THR group completed their weekly activities while mounted on a horse, whereas the BA control group completed their activities in a classroom within the riding facility.

The primary author attended sessions one and two for the THR group and sessions one, two, and eight for the BA control to conduct observations and interviews. Lessons for the THR and BA groups followed the same routine for each lesson: Both groups took place on a weekday afternoon, began with a five-minute art activity in a group activity room to allow for baseline data to be collected, followed by 45-minutes of activity based on the topic of the week, a 15-minute wrap-up activity, and a final five-minute art activity before the post-data was then collected.

THR Group

The THR group included 45 minutes of direct engagement on horseback and 15 minutes of grooming or tacking. A riding instructor led the group, and depending on the participants' needs, two to three volunteers supported each youth with their horse. The volunteers consisted of a horse lead, who held the reins of the horse, and one or two side walkers. Side walkers provided a variety of hands-on supports for individuals learning to ride by ensuring proper body positioning to prevent leaning or falling.

BA Group

The BA group included 45 minutes of classroom-based activities and 15 minutes of clean-up, utilizing a traditional classroom approach for instruction. Horses were occasionally visible through windows in the classroom and participants periodically walked around the THR site to view horses and equipment from a distance but had no direct interaction with horses. After 10 weeks of participation in the BA group, participants were given a free one-hour riding lesson.

Semi-structured Interviews

Two semi-structured interview guides were created to explore the experiences of these three young people. Given the age and disability status of the targeted sample, the IRB-approved interview guide consisted of four questions each; this allowed all interviews to be completed in no more than 30-minute increments. All three participants engaged in interviews one and two. The first interview was completed after the first session of the 10-week program and focused on gathering information about the participant's expectations of the program and building rapport. The second interview was completed after the tenth session and focused on gathering information about the experience of the program itself. After completing the 10-week program, the BA group participant completed a third interview after their free horseback riding lesson.

Youth were offered the option to allow a caregiver to assist them in their interview; their responses varied and are detailed within the results section. All interviews were recorded and transcribed verbatim by the primary author. A visual aid, Reproducible E (Kuypers & Sautter, 2021), from the Zones of Regulation™ was utilized during interviews to confirm answers that consisted of emotional vocabulary. The Zones of Regulation™ is a social-emotional curriculum developed to enhance emotional regulation and prosocial behaviors in Autistic youth (Kuypers, 2008). The Reproducible E visual aid uses words, pictures, and colors to allow multiple means of expression to confirm how an emotion was internally experienced and has been paired with animal-assisted research involving Autistic participants in the literature (Peters et al., 2024).

Ethical Considerations

This study included the interviewing of Autistic youth with and without caregivers present. The primary author who conducted the interviews was a licensed clinical social worker with over 10 years of experience working with Autistic youth as a mental health professional. The credentials and expertise of the interviewer supported the ethical application of this study protocol.

Analysis

In accordance with best practices in qualitative research (Tracy, 2010), consistent use of reflexivity (Cho & Trent, 2006) and congruent application of theory were important throughout the project (Carter & Little, 2007). Credibility was built using process member checking between participants and prolonged engagement between the research participants and the primary author (Morse, 2015).

Coding for this study was guided by Saldana's (2016) *Coding Manual for Qualitative Researchers*. Initial coding was conducted using in-vivo coding to remain close to the participant's own words. Second cycle coding employed axial coding to further categorize and compare codes within and between participants. A research assistant was trained in the coding protocol to increase the validity of the analysis and decrease any influence from researcher bias. The second coder reviewed the transcripts independently of the primary author. Codes were compared after each cycle of coding and discussed by both coders until consensus was established.

After coding, analysis followed a comparative case study approach by first focusing on the individual case studies and then searching for patterns between the three young people to explore shared meaning and lived experiences (Yin, 2018). A qualitative focusing strategy called the study trinity (Saldaña, 2016) was employed to organize codes into three main intersecting concepts that defined the experience. The study trinity approach organizes codes into three concepts defined by micro (individual), mezzo (organizational), and macro (societal) themes.

Results

To protect participant identity, yet respect the humanity of their lived experience, each participant has been assigned a pseudonym that will be used throughout this paper. The study trinity produced three concepts, one at each level. At the micro, or individual, level, the theme of voice emerged. This included codes on how comfortable the children felt using their voices and observations on how the youth used their voices to speak on their own behalf. Then, at the mezzo, or organizational, level, the theme of individualization emerged. This theme emerged from codes that expressed how the youth observed the adults modify and accommodate their needs and what these experiences felt like to the youth. Meanwhile, at the macro, or societal, level, the theme of experiencing difference emerged. Each youth shared a variety of ways in which past experiences with individualization, or a lack thereof, impacted their self-image and experience of disability.

Adam

Adam was a six-year-old boy; his first interview with the primary author was fourteen minutes and three seconds without a caregiver present. During this interview, which took place at the riding center, Adam requested to switch roles with the interviewer and engaged with available distractions such as the office telephone. Off-topic interactions were coded as distractions within the data, but also provide a glimpse into this young man's personality and expectations of adult interactions. Adam's second interview was completed with his mother present, at his request. This interview was 7 minutes and 56 seconds in length and took place in a hospital therapy office with minimal décor and furnishings to reduce distractions.

While the interviews were short in length, Adam was able to articulate his experience in the riding program and confirm those experiences through multiple means of expression during these interviews. He reported he felt a "big happy" about being in the riding program as he spun his body in circles, confirming his positive feelings through words and body movement. The interviewer also utilized a feelings rating scale (Kuypers & Sautter, 2021) to confirm the meaning behind the emotional language utilized by Adam during both interviews.

Brian

Brian was a 12-year-old boy who expressed himself verbally and met with the interviewer independently for both interviews. His first interview took fourteen minutes and twenty seconds and ended when Brian expressed a desire to switch topics and discuss a preferred interest, video games. His first interview was completed at the riding center, in an office with a closed door. The second interview was completed in a hospital therapy office and lasted sixteen minutes and forty-four seconds. There were notable differences in language usage between the first and second interviews. For example, Brian used the phrase "you know" fifteen times during his second interview. This linguistic pattern was not evident during his first interview.

Carrie

Carrie was a soft-spoken, sixteen-year-old girl who was interviewed three times to share her lived experience of the ten-week program and her post-intervention riding session. Her first interview was completed at the riding center with a caregiver present, at her request, and lasted thirty-three minutes and twenty seconds. Her second interview was completed in a hospital therapy office with the same caregiver present and lasted fourteen minutes and forty-six seconds. Her final interview was completed at the riding center after her free riding lesson. The interviewer was able to observe her entire riding session and then completed a five-minute and thirty-five second interview; Carrie's caregiver was not present during the third interview.

There are some notable differences between Carrie and the other participants. One difference was Carrie reported that her goal in this program was to try something that scared her. Unlike the two male participants in the trial group, Carrie reported feeling nervous about "riding them, slash like, being around them" when discussing her initial feelings about the riding program. When she did have the opportunity to ride a horse after the ten weeks of intervention, Carrie was hesitant to mount the horse. During her brief ride, Carrie was observed with tears rolling down her face, but in her third interview expressed, "I was kind of nervous at first, but then it started to get better, I tried to get more comfortable".

Cross Case Comparison

A cross-case comparison provides insight into the original research question about what these youths experienced and how it affected their lives. These comparisons sought to identify themes between and within the experiences of the youth participants as replications of experience, not as a sample of the population (Yin, 2018). The comparison found interdependent themes of voice, individualization, and difference, which were experienced in different ways by each participant.

Voice

All three youth used their voices differently. Adam and Carrie required support from adults, but for very different purposes. For example, Adam's first interview required a debrief meeting with his caregiver to confirm information that he had shared. Adam's speech and language differences made it difficult for the interviewer to fully comprehend his words without previous knowledge of his speech patterns. Carrie utilized her caregiver to support her self-advocacy. Carrie would process her experiences in the program with her caregivers at home; then her caregivers would share this feedback with the program staff on her behalf.

Meanwhile, Brian's demonstration of self-advocacy skills was independent. In our first interview, he reported,

> *"I went in with low expectations and, at first, it was hard, because it felt like they were, uh, babying me, but, uh, that was just for the first part and once I got that over with by basically telling them to be quiet, everything went great".*

Brian expressed no concerns about letting the adults around him know exactly what he felt. In fact, he expressed confidence in both interviews, even saying that he felt pride in being Autistic.

Brian and Adam also mentioned feeling like their horses listened to them, highlighting the positive experience of feeling heard. This was the THR participant's shared experience within the program. Brian provided insight in the following excerpt,

> *Brian: It's just because the horse is calm, then I get calm. I guess it feels nice to have somebody who actually listened to me, like, fully. You know?*
> *Interviewer: Tell me more about that.*
> *Brian: I finally... 'cause as a child, nobody really listens to me.*
> *Interviewer: You told me a lot about that, yea.*
> *Brian: A horse does. Undeniably.*

Individualization

Each child's voice exists on the micro level, but how they use their voice in a group setting moves this analysis into the mezzo level. The riding program was developed specifically for Autistic youth. While it was designed to be universally accessible for children with ASD level 1-3, the experience of ASD is unique to every individual. Adam, Brian, and Carrie each received accommodations, or individualization, through different applications of voice in the group setting.

Adam's use of voice to advocate on his own behalf was not coded in the transcripts. Based on his interviews, he "did all of it" and did not mention any issues accessing the THR program. Meanwhile, Brian reflected on his preference to interact with only the horse. He expressed awareness of his inflexible thinking and how this can interfere with developing human relationships. When asked to compare his experience bonding with his horse to building human relationships, he remarked, "like I'm worse to people because why should I be flexible to you even if I don't know you. You've never done anything, you know?" Brian's application of his independent and confident voice was not seeking support but instead asking to be left alone. This was in stark contrast to Carrie, who revealed in her interviews that she utilized her caregiver to seek increased challenge in her activities. Carrie shared that the adults working with her "… tried to find things that I would enjoy more and, like, make things harder for me", but this was only able to happen after it was brought to their attention that Carrie was feeling "childish".

Experience of Difference

The third theme was the experience of difference. This theme was created to code experiences in which the participants felt marginalized or "othered" due to a part of their identity. Participants expressed unique experiences with their identities in society that affected their perceptions and experiences within the THR and BA groups. Adam did not have any codes specific to this theme. Brian, however, had a lot of thoughts about his experience of feeling different. Contrary to the research question, Brian expressed that his age, not his disability, was a main contributor to feeling excluded. According to Brian, he prefers time spent with adults "because they're not as crazy" but he also does not always feel heard by adults.

Meanwhile, Carrie's interviews revealed a teenager in search of her place in society. Carrie and her caregiver shared a story about a previous experience in which Carrie felt different in another group setting. In that group, Carrie was the only Autistic participant and the only participant to not pass the class. Having been recently diagnosed, this was Carrie's first time participating in a program with other Autistic youth, but she still expressed a feeling that she did not fit in with the group.

Discussion

The results of this study create more questions than answers to the original research question: what are the experiences of Autistic youth who engage in THR, and how do those experiences affect their lives? Each of the participants' experiences was unique and highlighted the interconnectedness of micro, mezzo, and macro systems. For these three young people, their experience of THR was influenced by their personality, past lived experiences, and communication preferences. Meanwhile, the latter part of the research question, how this experience affected their lives, is difficult to ascertain. For example, Carrie expressed no interest in THR in the future, but will she try another program for Autistic youth? Her polite responses throughout the interview process make it difficult to fully understand the impact of this experience on her or hypothesize the lasting impact of another social rejection.

These three interviews highlight the importance of creating safe spaces for youth to express themselves. As the first qualitative study to interview Autistic youth about their experiences in a THR program, researchers can use these lived experiences to support the growing body of quantitative work, helping to enhance the trustworthiness and validity of collected data (Denzin, 2010). Currently, first-hand, lived experience is absent in literature.

The results of these case studies are congruent with emerging theories on how and why AAI work. For example, Temple Grandin hypothesizes that communication between Autistic people and animals has commonalities that contribute to the positive outcomes associated with AAI (Grandin et al., 2015). Adam and Brian both felt heard by the horse, and this deserves further probing in future research efforts. These two participants were participants in the THR group, and both endorsed the horse listening as being what they enjoyed the most. Unconditional positive regard has also been linked as a potential mediator for the experiences of AAI (Grandin et al., 2015). Youth in this project did not reflect feeling a nonjudgmental relationship with their horse directly, but the importance of relationship development and feeling heard is closely related. Future studies should consider exploring this aspect of AAI directly.

While this project does not confirm Dr. Grandin's suggestion that vestibular stimulation may be playing a role in why THR elicits social and emotional benefits without the application of a therapeutic framework (Grandin, et al., 2015), it does continue to support the hypothesis that something is happening during these interactions between animals and Autistic individuals. Additional qualitative work exploring the lived experience of Autistic youth in AAI with different species of animals could provide more insight into how and why such a non-therapeutic activity has therapeutic benefits. Furthermore, comparing experiences between THR and therapeutic EAS, like equine-assisted psychotherapy, may also provide insight into the science behind the human-animal bond and the outcomes associated with AAI.

Finally, these case studies are a reminder of the importance of ecological systems as originally described by Bronfenbrenner (1986), which situates the human experience through interconnected systems that influence our perception and complicate data collection in the social sciences. For any social-emotional wellness or mental health treatment program to meet expectations, we must remember that participants in these programs exist outside of the research setting. Even the most rigorous study cannot account for every variable affecting human experience. No single program can be the best fit for an entire population, particularly with a population as unique and diverse as the Autistic community. Further study of this ten-week manualized program will continue to identify information about who benefits from THR and what qualities in the people and animals are correlated with better outcomes.

There are several limitations to this study. The data collected in this study is the opinion of these three children. While their experiences are valid and significant, it is also important to recognize that more high-quality, qualitative research is needed. Future studies should aim to expand the sample size and consider additional interview opportunities. For example, a six-month follow-up interview after initial analysis would provide further insight and participant checking opportunities. Additionally, current trends in research involving Autistic participants suggest that the use of Autistic co-interviewers may have insider benefits (Kaplan-Kahn & Caplan, 2023). In this study, the interviewer had mixed insider/outsider status as a neurodivergent, allistic (non-Autistic) adult.

Conclusion

In relation to the research question, "what are the experiences of individuals with ASD who engage in AAI, and how do those experiences impact their lives?", an ecological system's view and use of the study trinity focusing strategy showed three themes that impact these experiences: voice, individualization, and experiences of difference. Given the limitations of case study research, the results of this investigation should be interpreted with caution, but they do remind us that not all Autistic children benefit from programs that cater specifically to Autism. Universal design may reduce barriers, but individual accommodations should always be made available.

Acknowledgments: The authors of this paper would like to acknowledge the contributions of Dr. Robin Gabriels, PsyD, who mentored and supported the primary author throughout this project. This project was completed as part of the primary author's dissertation, and we extend gratitude towards the members of the dissertation committee, including Dr. Helen Holmquist Johnson, Dr. Anne Williford, Dr. Tara Opsal, and Dr. Samantha Brown. Special thanks are extended to the youth and their families who participated in this project, as this would not have been possible without them.

Data Availability Statement: A full audit trail of the original data is available within the first author's dissertation. Banks, T. N. (2023). *Nothing about us, without us: Elevating voices from the autistic community* (Order No. 30530814). Available from ProQuest Dissertations & Theses Global. (2852326773)

Conflict of Interest: No potential conflicts of interest were reported by the authors.

Funds: No funds received.

Author Contributions: TB conceptualized and designed this study, conducted all interviews and transcription, and wrote the original draft of this manuscript. NK reviewed and edited the final manuscript. JH served as a second coder to validate analysis.

Statement: The study was conducted according to the guidelines of the Declaration of Helsinki and approved by the Institutional Review Board of Colorado State University (#3500) and University of Maryland Baltimore County (#1399).

References

American Psychiatric Association. (2013). *Diagnostic and statistical manual of mental disorders* (Fifth Edition). American Psychiatric Association. https://doi.org/10.1176/appi.books.9780890425596

American Veterinary Medical Association. (2025). *Animal-assisted interventions: Definitions*. AVMA Policies. https://www.avma.org/resources-tools/avma-policies/animal-assisted-interventions-definitions

Belardinelli, C., Raza, M., & Taneli, T. (2016). Comorbid behavioral problems and psychiatric disorders in autism spectrum disorders. *Journal of Childhood & Developmental Disorders, 02*(02). https://doi.org/10.4172/2472-1786.100019

Bölte, S., Holtmann, M., & Poustka, F. (2008). The social communication questionnaire (SCQ) as a screener for autism spectrum disorders: Additional evidence and cross-cultural validity. *Journal of the American Academy of Child & Adolescent Psychiatry, 47*(6), 719–720. https://doi.org/10.1097/CHI.0b013e31816c42bd

Bronfenbrenner, U. (1986). Ecology of the family as a context for human development: Research perspectives. *Developmental Psychology, 22*(6), 723–742. https://doi.org/10.1037/0012-1649.22.6.723

Buck, T. R., Viskochil, J., Farley, M., Coon, H., McMahon, W. M., Morgan, J., & Bilder, D. A. (2014). Psychiatric comorbidity and medication use in adults with autism spectrum disorder. *Journal of Autism and Developmental Disorders, 44*(12), 3063–3071. https://doi.org/10.1007/s10803-014-2170-2

Burnside, L., & Gabriels, R. (2019). *Barn Activity Group Manual* [Unpublished Manuscript].

Carter, S. M., & Little, M. (2007). Justifying knowledge, justifying method, taking action: Epistemologies, methodologies, and methods in qualitative research. *Qualitative Health Research, 17*(10), 1316–1328. https://doi.org/10.1177/1049732307306927

Cho, J., & Trent, A. (2006). Validity in qualitative research revisited. *Qualitative Research, 6*(3), 319–340. https://doi.org/10.1177/1468794106065006

Coman, D. C., Bass, M. P., Alessandri, M., Ghilain, C. S., & Llabre, M. M. (2018). Effect of equine-assisted activities on social and sensory functioning of children with autism. *Society & Animals, 26*(6), 551–575. https://doi.org/10.1163/15685306-12341479

Corbett, B. A., Bales, K. L., Swain, D., Sanders, K., Weinstein, T. A. R., & Muglia, L. J. (2016). Comparing oxytocin and cortisol regulation in a double-blind, placebo-controlled, hydrocortisone challenge pilot study in children with autism and typical development. *Journal of Neurodevelopmental Disorders, 8*(1), 32. https://doi.org/10.1186/s11689-016-9165-6

Corbett, B. A., Schupp, C. W., Levine, S., & Mendoza, S. (2009). Comparing cortisol, stress, and sensory sensitivity in children with autism. *Autism Research, 2*(1), 39–49. https://doi.org/10.1002/aur.64

Corbett, B. A., Schupp, C. W., Simon, D., Ryan, N., & Mendoza, S. (2010). Elevated cortisol during play is associated with age and social engagement in children with autism. *Molecular Autism, 1*(1), 13. https://doi.org/10.1186/2040-2392-1-13

Davis, T. N., Scalzo, R., Butler, E., Stauffer, M., Farah, Y. N., Perez, S., Mainor, K., Clark, C., Miller, S., Kobylecky, A., & Coviello, L. (2015). Animal assisted interventions for children with autism spectrum disorder: A systematic review. *Education and Training in Autism and Developmental Disabilities, 50*(3), 316–329.

Denzin, N. K. (2010). Moments, mixed methods, and paradigm dialogs. *Qualitative Inquiry, 16*(6), 419–427. https://doi.org/10.1177/1077800410364608

Gabriels, R. L., Agnew, J. A., Holt, K. D., Shoffner, A., Zhaoxing, P., Ruzzano, S., Clayton, G. H., & Mesibov, G. (2012). Pilot study measuring the effects of therapeutic horseback riding on school-age children and adolescents with autism spectrum disorders. *Research in Autism Spectrum Disorders, 6*(2), 578–588. https://doi.org/10.1016/j.rasd.2011.09.007

Gabriels, R. L., Pan, Z., Dechant, B., Agnew, J. A., Brim, N., & Mesibov, G. (2015). Randomized controlled trial of therapeutic horseback riding in children and adolescents with autism spectrum disorder. *Journal of the American Academy of Child & Adolescent Psychiatry, 54*(7), 541–549. https://doi.org/10.1016/j.jaac.2015.04.007

Gabriels, R. L., Pan, Z., Guérin, N. A., Dechant, B., & Mesibov, G. (2018). Long-term effect of therapeutic horseback riding in youth with autism spectrum disorder: A randomized trial. *Frontiers in Veterinary Science, 5*, 156. https://doi.org/10.3389/fvets.2018.00156

Grandin, T., Fine, A. H., O'Haire, M. E., Carlisle, G., & Bowers, C. M. (2015). The roles of animals for individuals with autism spectrum disorder. In A. Fine (Ed), *Handbook on animal-assisted therapy* (pp. 225–236) Elsevier. https://doi.org/10.1016/B978-0-12-801292-5.00016-X

Grandin, T., & Johnson, C. (2006). *Animals in translation: Using the mysteries of autism to decode animal behavior* (1. Harvest ed). Harcourt.

Hoagwood, K. E., Acri, M., Morrissey, M., & Peth-Pierce, R. (2017). Animal-assisted therapies for youth with or at risk for mental health problems: A systematic review. *Applied Developmental Science, 21*(1), 1–13. https://doi.org/10.1080/10888691.2015.1134267

Holm, M. B., Baird, J. M., Kim, Y. J., Rajora, K. B., D'Silva, D., Podolinsky, L., Mazefsky, C., & Minshew, N. (2014). Therapeutic horseback riding outcomes of parent-identified goals for children with autism spectrum disorder: An aba′ multiple case design examining dosing and generalization to the home and community. *Journal of Autism and Developmental Disorders, 44*(4), 937–947. https://doi.org/10.1007/s10803-013-1949-x

Kaplan-Kahn, E. A., & Caplan, R. (2023). Combating stigma in autism research through centering autistic voices: A co-interview guide for qualitative research. *Frontiers in Psychiatry, 14*, 1248247. https://doi.org/10.3389/fpsyt.2023.1248247

Kemeny, B., Burk, S., Hutchins, D., & Gramlich, C. (2022). Therapeutic riding or mindfulness: Comparative effectiveness of two recreational therapy interventions for adolescents with autism. *Journal of Autism and Developmental Disorders, 52*(6), 2438–2462. https://doi.org/10.1007/s10803-021-05136-z

Kuypers, L. M. (2008). *A curriculum designed to foster self-regulation in students with neurobiological impairments.* [Capstone, Hamline University]. https://digitalcommons.hamline.edu/hse_all/2124

Kuypers, L. M., & Sautter, E. (2021). *Reproducible E - Elementary ages.* Think Social Publishing Inc. https://www.socialthinking.com/zones-of-regulation/free-stuff

Lord, C., Rutter, M., DiLavore, P., Risi, S., Gotham, K., & Bishop, S. (2012). *Autism Diagnostic Observation Schedule: ADOS-2.* Western Psychological Services. https://www.wpspublish.com/ados-2-autism-diagnostic-observation-schedule-second-edition

Lovrić, R., Farčić, N., Mikšić, Š., & Gvozdanović Debeljak, A. (2020). Parental perception of changes in basic life needs of children with disabilities after six months of therapeutic horseback riding: A qualitative study. *International Journal of Environmental Research and Public Health, 17*(4). https://doi.org/10.3390/ijerph17041213

Maenner, M. J., Warren, Z., Williams, A. R., Amoakohene, E., Bakian, A. V., Bilder, D. A., Durkin, M. S., Fitzgerald, R. T., Furnier, S. M., Hughes, M. M., Ladd-Acosta, C. M., McArthur, D., Pas, E. T., Salinas, A., Vehorn, A., Williams, S., Esler, A., Grzybowski, A., Hall-Lande, J., … Shaw, K. A. (2023). Prevalence and characteristics of autism spectrum disorder among children aged 8 years—Autism and Developmental Disabilities Monitoring Network, 11 Sites, United States, 2020. *MMWR. Surveillance Summaries, 72*(2), 1–14. https://doi.org/10.15585/mmwr.ss7202a1

Merritt, T., & Gabriels, R. (2020). *Therapeutic horseback riding manual (2nd Edition) for children with autism spectrum disorder* [Unpublished Manuscript].

Morse, J. M. (2015). Critical analysis of strategies for determining rigor in qualitative inquiry. *Qualitative Health Research, 25*(9), 1212–1222. https://doi.org/10.1177/1049732315588501

Nieforth, L. O., Schwichtenberg, A. J., & O'Haire, M. E. (2021). Animal-assisted interventions for autism spectrum disorder: A systematic review of the literature from 2016 to 2020. *Review Journal of Autism and Developmental Disorders, 10*, 255-280. https://doi.org/10.1007/s40489-021-00291-6

O'Haire, M. E. (2013). Animal-assisted intervention for autism spectrum disorder: A systematic literature review. *Journal of Autism and Developmental Disorders, 43*(7), 1606–1622. https://doi.org/10.1007/s10803-012-1707-5

Pan, Z., Granger, D. A., Guérin, N. A., Shoffner, A., & Gabriels, R. L. (2019). Replication pilot trial of therapeutic horseback riding and cortisol collection with children on the autism spectrum. *Frontiers in Veterinary Science, 5*, 312. https://doi.org/10.3389/fvets.2018.00312

Peters, B. C., Gabriels, R. L., Schmid, A. A., Pan, Z., Merritt, T., Hoffman, A., & Hepburn, S. (2024). The feasibility of occupational therapy using Zones of Regulation™ Concepts in an equine environment. *Frontiers in Psychiatry*, *15*, 1401222. https://doi.org/10.3389/fpsyt.2024.1401222

Professional Association of Therapeutic Horsemanship (PATH) International. (2023). *Equine-assisted services: Definition guide for researchers*. https://pathintl.org/wp-content/uploads/2023/07/PATH-Intl-Definitions-Guide-for-Researchers.pdf

Protopopova, A., Matter, A. L., Harris, B. N., Wiskow, K. M., & Donaldson, J. M. (2019). Comparison of contingent and noncontingent access to therapy dogs during academic tasks in children with autism spectrum disorder. *Journal of Applied Behavior Analysis*, *53*(2), 811-834. https://doi.org/10.1002/jaba.619

Rodriguez, K. E., Guérin, N. A., Gabriels, R. L., Serpell, J. A., Schreiner, P. J., & O'Haire, M. E. (2018). The state of assessment in human-animal interaction research. *Human-Animal Interaction Bulletin*, *6*(Special Edition), 63–81.

Rosen, T. E., Mazefsky, C. A., Vasa, R. A., & Lerner, M. D. (2018). Co-occurring psychiatric conditions in autism spectrum disorder. *International Review of Psychiatry*, *30*(1), 40–61. https://doi.org/10.1080/09540261.2018.1450229

Saldaña, J. (2016). *The coding manual for qualitative researchers* (3E [Third edition]). SAGE.

Tracy, S. J. (2010). Qualitative quality: Eight "big-tent" criteria for excellent qualitative research. *Qualitative Inquiry*, *16*(10), 837–851. https://doi.org/10.1177/1077800410383121

Ward, S. C., Whalon, K., Rusnak, K., Wendell, K., & Paschall, N. (2013). The association between therapeutic horseback riding and the social communication and sensory reactions of children with autism. *Journal of Autism and Developmental Disorders*, *43*(9), 2190–2198. https://doi.org/10.1007/s10803-013-1773-3

Yin, R. K. (2018). *Case study research and applications: Design and methods* (Sixth edition). SAGE.

Zoccante, L., Marconi, M., Ciceri, M. L., Gagliardoni, S., Gozzi, L. A., Sabaini, S., Di Gennaro, G., & Colizzi, M. (2021). Effectiveness of Equine-Assisted Activities and Therapies for improving adaptive behavior and motor function in Autism Spectrum Disorder. *Journal of Clinical Medicine*, *10*(8), Article 8. https://doi.org/10.3390/jcm10081726

The effects of a short-term intensive hippotherapy on a child with cerebral palsy: A case study

Maria Liga, MSc[1], Marina Pantiera, MSc[2], Antigoni Kaftantzi, MSc[1], Alexandra Stergiou, PhD[1], Filippos Krontiras[1], Alexandra Hristara, Professor[2], Anna Halkia, Professor[2] Ilias Kallistratos, Professor[2],

[1]Therapeutic Riding Center Serres (KETHIS), Serres, Greece, [2]International Hellenic University (DIPAE), Thessaloniki, Greece

Research has shown that hippotherapy is effective in improving gross motor function in individuals with cerebral palsy. In addition, it has shown positive effects in restoring balance and reducing muscle spasticity. The aim of this study was to evaluate the effects of a short-term, intensive hippotherapy program consisting of daily sessions over six consecutive days on the functional level of a child with cerebral palsy. The study included a child with cerebral palsy (spastic quadriplegia) aged 3.5 years. To evaluate the functional level of the child, gross mobility was measured using the GMFM-88 and GREAT scales, which evaluate the posture of the body and head on the horse. The physiotherapy intervention included ten sessions of hippotherapy. The child was initially evaluated one day before the intervention and re-evaluated one day after the end of the sessions. Measurements taken before and after the intervention indicated improvements in gross mobility and posture of the child's body and head according to the evaluation scales. It was concluded that an intensive hippotherapy program may improve gross motor function in children with cerebral palsy, as measured using the GMFM-88 and GREAT scales. Further, this experience may be positive and motivating for children. However, the conclusions of this study cannot be considered reliable and generalizable because the sample consisted of only one child. Further research should be performed on a larger sample of children with cerebral palsy.

All correspondence and requests for reprints should be addressed to: Maria Liga, at ligamaria2006@yahoo.gr or 1, A. Argirou, 62100, Serres, Greece.

Handling Editor: Ines Pereira-Figueiredo
Reviewers: Gunilla Silfverberg & Laura McFarland

Volume 25, Number 1, January 2026

DOI:

Keywords: cerebral palsy, equine-facilitated physiotherapy, hippotherapy, gross motor function, GMFM-88, GREAT

Cerebral Palsy (CP) is a group of permanent, but not irreversible, disorders of movement and/or posture and motor function resulting from non-progressive interference, damage, or abnormality of the developing, immature brain (Bax et al., 2005; Cans et al., 2007; Rosenbaum et al., 2007), and faulty transmission of signals between the central nervous system and muscles. This imbalance leads to increased muscle activity, which affects posture and gait patterns, which in turn results in limitations in daily activities and independence (Morris, 2007).

Physiotherapy deals with individuals, adults, and children who present with neurological, musculoskeletal, sensory, and psychosocial disorders. Recently, interest has turned to alternative forms of therapy for the treatment of these disorders that complement the role of physiotherapy and are engaging and creative. Hippotherapy is one such alternative form of therapy.

Hippotherapy or equine-assisted physiotherapy promotes, supports, and restores physical health through interaction with horses, delivered exclusively by accredited health professionals who hold specialized certification in this form of treatment. In this article, the term "hippotherapy" is preferred over "equine-assisted physiotherapy", as it derives from the Greek word "hippos" (horse), reflecting the historical roots of the therapeutic application of horse movement in Greek antiquity, where its therapeutic value has long been recognized. During hippotherapy, the three-dimensional movement of the horse's hips and pelvis presents a movement challenge to the rider (Silkwood-Sherer et al., 2012). The horse's gait creates a rhythmic motion that not only prompts the rider's forward and backward swinging movements, but also offers sensory stimulation through its varied, rhythmic, and repetitive actions. This horse movement closely resembles the natural motion of the human pelvis during walking. During the session, the therapist engages the horse's characteristic movements to provide carefully graded motor and sensory inputs to the patient. In addition, they help develop control of the head and trunk posture and improve endurance, symmetry, awareness, and perception of the body (Meregillano, 2004). Thus, a foundation is created for improving neurological function and sensory processing, which can be generalized to a wide range of daily activities and have functional outcomes.

Another important parameter is the transmission of warmth from the horse's back to the patient's body. This mechanism of action is important for normalizing muscle tone, reducing spasticity, and increasing muscle elasticity. These abilities are essential for an individual to perform movements that involve a wide range of joints. Many studies have demonstrated favorable physical effects of hippotherapy on coordination, muscle tone, stiffness, flexibility, endurance, strength, correction of abnormal motor patterns, and gait improvement (Meregillano, 2004; Shurtleff et al., 2009; Shurtleff & Engsberg, 2010; Silkwood-Sherer et al., 2012; Silkwood-Sherer & Warmbier, 2007; Stergiou et al., 2017). Numerous studies have confirmed the beneficial effects of hippotherapy on postural control, balance, gross motor function, and functional performance in children with CP (Kwon et al., 2015; Moraes et al., 2016; Tseng et al., 2012; Zadnikar & Kastrin, 2011).

The interaction between humans and horses is the basis of therapy due to the emotional responses of humans to horses. Motivated by this emotional response and the joy that a horse can provide, hippotherapy serves as an excellent treatment for children with disabilities, including those with CP (Zadnikar & Kastrin, 2011).

Although numerous studies have demonstrated the benefits of hippotherapy in children with cerebral palsy, few have explored the outcomes of intensive, short-term programs that deliver multiple sessions over a short period. The present case study sought to address this gap by examining the effects of an intensive hippotherapy program on a child with spastic quadriplegia.

The present study focuses on a short-term, intensive hippotherapy program implemented for a young child with spastic quadriplegia who exhibited no cognitive or emotional impairments and demonstrated excellent perceptual abilities. This study aimed to assess the effects of a short-term, intensive hippotherapy intervention on gross motor function and posture in children with cerebral palsy.

Method

The hippotherapy program was conducted at the Serres Therapeutic Riding Center (KETHIS) in Neos Skopos, Greece. The sessions were performed mostly in an indoor arena and some outside in an open-air area.

Written informed consent was obtained from the child's legal guardians before participation. The guardians were informed of the study objectives, procedures, potential benefits, and risks. Participation was voluntary and the guardians had the right to withdraw the child from the study at any point without any negative consequences. All data were handled confidentially, and the child's identity was anonymized in all reports and publications. This study did not receive formal institutional ethics committee approval as it was conducted as part of routine clinical care/service evaluation. All procedures adhered to ethical standards to protect the participant's rights and confidentiality.

The horses were led by experienced, specially trained instructors. Two physiotherapists were positioned on either side of the horse to deliver therapeutic interventions throughout each session while ensuring the child's safety during activities.

In the present study, "intensive" hippotherapy program refers to a short-term intervention, consisting of ten sessions over six consecutive days, starting with 10-minute sessions that progressively extended to 20 minutes. During the first three days, two sessions were held daily at 10:00 am and 4:00 pm. On the fourth day, only an afternoon session was conducted. The fifth day included two sessions again, and on the sixth and final day, only a morning session took place. The hippotherapy intervention protocol was tailored to the needs of a 3.5-year-old child with spastic quadriplegia. Sessions included structured exercises targeting posture, trunk control, balance, and coordination. Therapists closely monitored the child's responses during each session, including muscle tone, attention, engagement, and signs of fatigue, such as restlessness, reduced ability to maintain posture, or decreased participation. Exercises were adapted in real time to accommodate these observations, either by modifying the activity, reducing its duration, or incorporating rest periods, ensuring safety and maximizing therapeutic benefits.

Three horses, with smooth, rhythmic movement, no abrupt changes in tempo, and specially trained for therapeutic purposes, participated throughout the sessions. The choice of horse and specific movements performed were individualized based on the child's responses and physical needs. Because the child was young and required proper hip abduction, a short-statured horse with a narrow back was selected for some sessions to allow the child to sit comfortably astride without excessive leg abduction and maintain correct leg positioning during hippotherapy. Additionally, the therapist experimented with different horse movements and the sensory stimuli they generated to observe their effects on various body positions of the child on horseback. During the initial sessions, the child's compatibility with each horse was systematically assessed using standardized clinical criteria including postural stability, trunk control, muscle tone, emotional response, engagement, and overall comfort. Behavioral indicators, such as willingness to participate, facial expressions, and ease of following instructions were also evaluated.

Based on these comprehensive observations, the horse that elicited the most positive functional and emotional responses, as demonstrated by greater postural stability, improved trunk control, reduced muscle tone abnormalities, and enhanced engagement, was selected as the primary mount for most ongoing sessions. However, to accommodate scheduling constraints, maintain the child's motivation, and ensure safety, some sessions involved the use of a tall horse of a large build and a medium-sized horse of a medium build. This variation was necessary due to horse availability and to provide diverse sensory inputs, although the majority of the ten sessions were conducted with the short-statured horse, 1.30 cm at the withers, that showed the best compatibility with the child.

Regarding the horse's back surface, some sessions were performed with a therapeutic pad to enhance comfort and reduce sensory overstimulation, while others were conducted on the horse's bare back to provide varied tactile experiences depending on the child's tolerance and therapeutic goals. Documenting these observations and adjustments ensures transparency in clinical decision-making and highlights the individualized and adaptive nature of hippotherapy interventions.

The hippotherapy protocol involved various positions for the child on the horse. The intervention was performed with respect to the child's response and fatigue level with respect to the horse itself.

A detailed description of the full intervention protocol including session structure, specific exercises, and progression is provided in Table 1.

Table 1

Structure and Adaptation of the Intensive Hippotherapy Intervention Sessions

Phase	Duration	Activities / Exercises	Adaptation / Monitoring
Warm-up familiarization with the horse and the environment	5 min	Gentle stretching of upper/lower limbs; attention-engaging games.	Adjust intensity and duration based on child's engagement and muscle tone.
Core Exercises Mounting and positioning	10 min	Posture & trunk control: sitting tall, reaching, maintaining balance- Balance & coordination: side-to-side, forward-backward movements, arm coordination. Gait simulation & muscle activation: slow walking, turning, controlled movements.	Monitor fatigue, muscle tone, attention; modify exercises or provide rest as needed.
Cool-down	5 min	Gentle stretching on/off horse; relaxation activities.	Observe relaxation response; adjust time or intensity if fatigue or tension persists.
Individualization	Entire session	Continuous observation of child's response (posture, engagement, fatigue). Horse selection based on stability, trunk control, and emotional response.	Exercises and horse movements adapted in real time. according to child's and horse's comfort and capabilities.
Progression	Across program	Gradual increase in exercise complexity and duration.	Weekly re-evaluation guides adjustments

	for safety and optimal functional gains.

The child was assessed using the Gross Motor Function Measure (GMFM-88) Gross Motor Skills Scale and the Gainesville Riding through equine-assisted therapy (GREAT) scale one day before the intervention and one day after the end of the sessions. During the six days, no other physiotherapy program was followed, except for hippotherapy.

Outcome Measures

GMFM

The Gross Motor Function Measure (GMFM) is a scale with specific criteria that indicate the functional ability of a child with CP. The gross motor skills of patients were evaluated based on the milestones of the normal development of gross motor skills (Russell et al., 2000). The GMFM-88 scale, which includes 88 tests, comprises five categories (Alotaibi et al., 2014). The tests concern various motor skills that a child can perform alone, without the help of another person. The tests were grouped into five categories of motor function.

A. Horizontal position and rolling (17 tests)
B. Sitting position (20 tests)
C. Crawling and kneeling position (14 tests)
D. Standing (13 tests)
E. Walking, running and jumping (24 tests)

Each category can be used as an independent module. Each GMFM test was scored on a four-point Likert scale (0-1-2-3). At zero (0), the child cannot initiate movement. At step (1), the movement is initiated, but less than 10% of the exercise is performed. At two (2), the test is partially performed (more than 10% and less than 100%), and at three (3), the child can perform and complete the movement. The score for each module is the sum of the scores for its individual activities, divided by the corresponding maximum score for the module, and multiplied by 100 to obtain a percentage (Russell et al., 2000).

According to general guidelines, the GMFM (Russell et al., 2000) should be applied in a comfortable room (e.g., a physiotherapy room or a large room), where the space meets the outlined requirements. These requirements mainly concern the equipment needed to perform the tests (toys, steps, etc.) and specific tools (stopwatch, stick). The child must be wearing light clothing and not wearing shoes, and the level of severity of their condition must have been determined according to the Gross Motor Function Classification System (GMFCS) scale (Palisano et al., 2008). In this study, children were examined in categories B, C, and D of the GMFM.

GREAT

The Gainesville Riding through equine-assisted therapy (GREAT) Scale (Frease, 1996) is an indicator of postural control and balance and includes five specific assessment areas: 1) head/cervical, 2) shoulder/thoracic spine, 3) pelvis/lumbar spine, 4) hip angle, and 5) knee flexion/heel orientation. The scores for the five areas examined were as follows.

For the head and cervical spine: score two (2) if the head extends in front of the shoulder and the cervical spine is flexed; one (1) if the head is upright and balanced over the torso but the cervical spine is flexed looking down; zero (0) if the head is upright and balanced over the torso with the ear aligned with the center of the shoulder; three (3) if the head extends in front of the shoulder and the cervical spine is also extended; four (4) if the head is balanced on the extended cervical spine, looking up (Frease, 1996).

For shoulder and thoracic spine: score two - moderate kyphosis with shoulder in front of hip and observing a continuous curve; score one - mild kyphosis with shoulder in front of hip and in a slump; zero, normal thoracic curve with shoulders aligned over hip; three-flattened thoracic curve with shoulders back; four-flattened thoracic curve with shoulders back but raised (Frease, 1996).

For the pelvis and lumbar spine: score two – moderate posterior pelvic tilt with blocked lumbar curve; one - mild posterior pelvic tilt with reduced lumbar lordosis; zero, neutral pelvic position with normal lumbar lordosis; three - mild anterior pelvic tilt with normal to mildly increased lumbar lordosis; four - moderate anterior pelvic tilt with increased lordosis.

For hip angle: score one - when the hip angle is open; zero - when the hip angle is approximately 135 degrees; three - when the hip angle is closed (Frease, 1996).

For knee flexion and heel orientation: score one - when the knee is extended with the heel down and can be in front or behind the pelvis; zero - when the knee is bent with the foot perpendicular to the ground on the stirrup and the heel aligned with the pelvis; three - when the knee angle is close to the heel pulled and can be in front or behind the pelvis (Frease, 1996).

The assessment of positions for this scale was performed from a side view through video recordings, while the child remained in a stable sitting position looking forward. The two physiotherapists who participated in the study jointly conducted the assessments and reached a consensus on each item during the scoring. Both raters were experienced in pediatric physiotherapy and familiar with the use of the GMFM-88 and GREAT scales. Although formal inter-rater reliability statistics were not calculated, the collaborative rating process ensured consistency in the evaluation of the child's performance. Future studies should include independent raters and report inter-rater reliability indices to enhance the methodological rigor. The physiotherapists analyzed the video recordings (i.e., before and after hippotherapy) and rated the recordings according to the provided scales. Scoring was completed on the last day of the tenth session. The scores of each domain were added together to produce a total score, which shows how close the child is to proper body alignment. A total score of 0 indicates excellent posture, meaning that scores above 0 indicate some postural problems (Frease, 1996).

Results

The results of the exercise tests for each of the three GMFM categories before and after the hippotherapy program are shown in Table 2.

Table 2

Pre- and Post-Intervention Scores on the GMFM-88 and G.R.E.A.T. Scales

Scale / Category	Pre-intervention Score	Post-intervention Score	Change
GMFM-88			
Category B (Sitting)	33%	67%	+34%
Category C (Crawling & Kneeling)	24%	50%	+26%
Category D (Standing)	10%	20%	+10%
Total (B + C + D)	22%	46%	+24%
G.R.E.A.T. Total Score	9	2	–7

The total score based on the G.R.E.A.T scale before the hippotherapy sessions was nine, and after the treatments ended it was two.

Minimally Clinically Important Difference (MCID)

According to the published guidelines for the GMFM-88 (Oeffinger et al., 2008; Russell et al., 2000), a change of approximately 5-7 percentage points represents the MCID for children with cerebral palsy. In the present case, the total score increased by 24 percentage points (from 22% to 46%), indicating a clinically meaningful improvement in the gross motor function. For the G.R.E.A.T. scale, no specific MCID values have been reported in the literature, but the decrease in total score from 9 to 2 reflects a substantial improvement in postural alignment and trunk control, consistent with the child's observed functional gains.

Discussion

This case study specifically examined an intensive short-term hippotherapy program designed to test whether concentrated exposure over a few days could lead to functional and postural improvements. Based on Table 2, it is clear that the child more than doubled their performance score after the intensive hippotherapy program. These results suggest that even a brief but intensive intervention can produce clinically meaningful gains in children with CP.

From a clinical perspective, hippotherapy has been used in patients with various disabilities, and the findings have shown that it represents a substantial intervention in neurological rehabilitation (Guindos-Sanchez et al., 2020; Matusiak-Wieczorek et al., 2020; Menor-Rodriguez et al., 2021; Moraes et al., 2016). This study contributes to the limited body of literature on intensive hippotherapy interventions in children with CP. The positive changes observed in gross motor function and postural control suggest that short-term high-frequency hippotherapy may yield measurable benefits.

In this study, according to the recordings and the two scales used, the intervention appeared to contribute to a positive outcome. However, few studies have attempted to intervene with hippotherapy by implementing a short-term intensive program. Using PubMed, Google Scholar, and PEDro, only two articles emerged that studied a dynamically intensive hippotherapy program for CP. One study that emerged from this search was Žalienė et al. (2018). This study evaluated the effects of horseback riding on beginners (short-term hippotherapy sessions) and advanced (long-term) riders with common mobility problems resulting from CP and affecting their entire mobility. The results of this study showed that ten hippotherapy sessions did not affect novice riders with CP, and their gross motor function and level of motor function remained unchanged. However, in half of the advanced riders with CP, the gross motor function improved significantly. Additionally, the level of gross motor skill performance in the four advanced riders increased, but this difference was not statistically significant.

Honkavaara and Rintala (2010) investigated the effects of short-term hippotherapy on functional gait changes in children with CP using an intensive hippotherapy program. The results of this study showed that it is possible that short-term hippotherapy could improve functional gait in children with CP.

The improvements in such studies are likely to be observed due to the non-clinical environment in which the treatment is carried out. The sessions were conducted at an equestrian center in a natural environment. In addition, contact with animals evokes pleasant emotions and encourages greater interaction and participation of the person during treatment. Therefore, children are mainly excited and actively participate in motor activities that are fun and simultaneously therapeutic, which is essential for clinical improvement and positive therapeutic results. Continuous interest in treatment, motivation, and pleasure is certain, all of which directly affect the results of the treatment. Because CP is a chronic condition that often requires long-term participation in therapeutic activities, it is important that hippotherapy exists as a complementary and alternative treatment that combines both rehabilitation and fun.

The results of this case study indicate that a short-term, intensive hippotherapy program can lead to meaningful functional improvements in a young child with spastic quadriplegia. The GMFM-88 score increased by 24 percentage points, well above the minimally clinically important difference of 5-7%, suggesting substantial gains in gross motor function. Similarly, a reduction in the GREAT scale score from 9 to 2 reflects improved postural alignment and trunk control. The improvements on this scale are seen from the lower total score and suggest that hippotherapy helped improve posture and trunk control for the child, which allowed further functional improvement. These results are most likely due to the constant movement of the horse, forcing the child to constantly correct their trunk positions to remain stable and upright on the horse. This process helps strengthen the trunk muscles, allowing for better postural control after treatment. All these findings support the value of individualized hippotherapy tailored to the child's responses and compatible with the selected horse. Although based on a single case and therefore not generalizable, this report highlights the potential of hippotherapy to enhance motor function and postural control in children with cerebral palsy and underscores the need for further research.

Hippotherapy provides continuous, rhythmic, three-dimensional movement through the horse's gait, stimulating the vestibular, proprioceptive, and tactile sensory systems (Meregillano, 2004). This multisensory input facilitates improved postural control and balance by enhancing sensory processing and integration, consistent with the sensory integration theory. Horse movement mimics natural pelvic motion during walking, promoting normalization of muscle tone and reducing spasticity via sensory-driven modulation (Meregillano, 2004; Silkwood-Sherer et al., 2012). Additionally, warmth transmitted from the horse's body contributes to muscle tone normalization and increased elasticity, supporting motor function (Meregillano, 2004).

The intervention in this study involved structured exercises targeting posture, trunk control, balance, and coordination, tailored to the child's individual needs and adapted in real-time based on fatigue and engagement. This approach aligns with task-specific training principles in which repetitive, goal-directed activities promote motor learning and functional gains (Alotaibi et al., 2014; Russell et al., 2000). The child's continuous postural adjustments required to maintain stability on the moving horse reinforce motor control and strength in functional tasks, contributing to the clinically meaningful improvements observed in the GMFM and GREAT scores.

Intensive, short-term programs deliver frequent, consistent motor and sensory challenges, which can stimulate neuroplastic changes by encouraging the reorganization of neural pathways involved in motor control and postural regulation (Matusiak-Wieczorek et al., 2020; Menor-Rodriguez et al., 2021). Emotional engagement and motivation elicited by interaction with the horse and the natural, non-clinical environment likely enhance attention and participation, further supporting neuroplasticity (Zadnikar & Kastrin, 2011). These mechanisms underline the substantial gains in gross motor function and postural alignment after only ten sessions.

Integrating these frameworks strengthens scientific interpretation by connecting the observed clinical outcomes to the underlying mechanisms. Hippotherapy's unique combination of multisensory stimulation, task-specific motor practice, and emotional engagement creates a therapeutic environment that is conducive to functional recovery in children with cerebral palsy. As research on intensive hippotherapy programs is extremely limited, more research is necessary to confirm the positive and catalytic role of hippotherapy as an intensive intervention for children with CP.

Although the results of this case study were encouraging, several limitations should be acknowledged. Interrater reliability was not formally assessed, as the two physiotherapists rated the video recordings together by consensus, which may have introduced bias or reduced the objectivity of the evaluation. Nevertheless, both raters were experienced in pediatric physiotherapy and were familiar with the GMFM-88 and GREAT scales, which likely supported consistent scoring. Additionally, the findings were based on a single participant, limiting the generalizability of the results to a broader population of children with cerebral palsy. The study also lacked a follow-up period to assess the durability and long-term effects of intensive hippotherapy. Given the extremely limited research on intensive hippotherapy programs, further studies with larger sample sizes, independent raters, and longitudinal designs are necessary to validate and extend these preliminary findings and confirm the positive and catalytic role of hippotherapy as an intensive intervention in children with CP.

Conclusions

In general, the body of research conducted over the years on the effects of hippotherapy in children with cerebral palsy has consistently reported positive results. Most studies have identified hippotherapy as a means of improving the acquisition of motor skills, mainly in pediatric patients with CP. The contribution of hippotherapy to posture, postural control, movement, endurance, coordination, balance, strength and psychosocial benefits is useful and has an important role. Therefore, the results of hippotherapy are multifunctional and not indicated in just a single benefit; rather, after hippotherapy, multiple positive changes were observed in the physical and psycho-emotional state of the rider.

The present study suggests that an intensive hippotherapy program may significantly improve the functional level of a child with CP, as well as the trunk and head posture. Improvement was observed, as evidenced by changes in the scores on the GMFM and the GREAT scales from before the intervention to immediately after. Since this is a case study, further investigation with subsequent studies with a larger sample size and control groups are necessary to validate the results.

Data Availability Statement: Data available on request from the corresponding author.

Conflict of Interest: The authors declare no conflicts of interest.

Funds: No funds received.

Author Contributions: Conceptualization, MP; methodology, AH, IK, AH; formal analysis, MP, AH, IK, AH; investigation, ML, MP, FK; resources, MP, AH, AS; data curation, ML, MP; writing-original draft preparation, MP, ML, AH.; writing-review and editing, IK, AH, AS; project administration ML. All authors have read and agreed to the published version of the manuscript.

Statement: This study was conducted in accordance with the guidelines of the Declaration of Helsinki.

References

Alotaibi, M., Long, T., Kennedy, E., & Bavishi, S. (2014). The efficacy of GMFM-88 and GMFM-66 to detect changes in gross motor function in children with cerebral palsy (CP): a literature review. *Disability and rehabilitation*, *36*(8), 617–627. https://doi.org/10.3109/09638288.2013.805820

Bax, M., Goldstein, M., Rosenbaum, P., Leviton, A., Paneth, N., Dan, B., Jacobsson, B., & Damiano, D. (2005). Proposed definition and classification of cerebral palsy, April 2005. *Developmental Medicine & Child Neurology*, 47(8), 571–576. https://doi.org/10.1017/s001216220500112x

Cans, C., Dolk, H., Platt, M., Colver, A., Prasauskene, A., & Rägeloh-Mann, I. K. (2007). Recommendations from the SCPE collaborative group for defining and classifying cerebral palsy. *Developmental Medicine & Child Neurology*, 49, 35–38. https://doi.org/10.1111/j.1469-8749.2007.tb12626.x

Frease, L. (1996). *The G.R.E.A.T. Postural Scale*. Gainesville, FL: University of Florida.

Guindos-Sanchez, L., Lucena-Anton, D., Moral-Munoz, J., Salazar, A., & Carmona-Barrientos, I. (2020). The effectiveness of hippotherapy to recover gross motor function in children with cerebral palsy: a systematic review and meta-analysis. *Children*, 7(9), 106. https://doi.org/10.3390/children7090106

Honkavaara, M., & Rintala, P. (2010) The influence of short term, intensive hippotherapy on gait in children with cerebral palsy. *European Journal of Adapted Physical Activity*, 3(2), 29-36.

Kwon, J., Chang, H. J., Yi, S., Lee, J. Y., Shin, H., & Kim, Y. (2015). Effect of hippotherapy on gross motor function in children with cerebral palsy: a randomized controlled trial. *The Journal of Alternative and Complementary Medicine*, 21(1), 15–21. https://doi.org/10.1089/acm.2014.0021

Matusiak-Wieczorek, E., Dziankowska-Zaborszczyk, E., Synder, M., & Borowski, A. (2020). The influence of hippotherapy on the body posture in a sitting position among children with cerebral palsy. *International Journal of Environmental Research and Public Health*, 17(18), 6846. https://doi.org/10.3390/ijerph17186846

Menor-Rodríguez, M. J., Martín, M. S., Sánchez-García, J. C., Montiel-Troya, M., Cortés-Martín, J., & Rodríguez-Blanque, R. (2021). Role and effects of hippotherapy in the treatment of children with cerebral palsy: a systematic review of the literature. *Journal of Clinical Medicine*, 10(12), 2589. https://doi.org/10.3390/jcm10122589

Meregillano G. (2004). Hippotherapy. *Physical Medicine and Rehabilitation Clinics of North America*, 15, 843–54.

Moraes, A. G., Copetti, F., Angelo, V. R., Chiavoloni, L. L., & David, A. C. (2016). The effects of hippotherapy on postural balance and functional ability in children with cerebral palsy. *Journal of Physical Therapy Science*, 28(8), 2220–2226. https://doi.org/10.1589/jpts.28.2220

Morris, C. (2007). Definition and classification of cerebral palsy: a historical perspective. *Developmental Medicine & Child Neurology*, 49, 3–7. https://doi.org/10.1111/j.1469-8749.2007.tb12609.x

Oeffinger, D. J., et al. (2008). "Comparison of GMFM-88 and GMFM-66 scores in children with cerebral palsy." *Physical Therapy*, 88(6), 629–639.

Palisano RJ, Rosenbaum P, Bartlett D, Livingston MH. (2008). Content validity of the expanded and revised Gross Motor Function Classification System. *Developmental Medicine and Child Neurology,* 50(10), 744-50. https://doi.org/10.1111/j.1469-8749.2008.03089.x

Rosenbaum, P., Paneth, N., Leviton, A., Goldstein, M., Bax, M., Damiano, D., Dan, B., & Jacobsson, B. (2007). A report: the definition and classification of cerebral palsy April 2006. *Developmental Medicine and Child Neurology. Supplement*, 109, 8–14.

Russell, D. J., Rosenbaum, P. L., Avery, L. M., & Lane, M. (2000). *Gross Motor Function Measure (GMFM-66 & GMFM-88) User's Manual.* London: Mac Keith Press.

Shurtleff, T. L., & Engsberg, J. R. (2010). Changes in trunk and head stability in children with cerebral palsy after hippotherapy: a pilot study. *Physical & Occupational Therapy in Pediatrics*, 30(2), 150–163. https://doi.org/10.3109/01942630903517223

Shurtleff, T. L., Standeven, J. W., & Engsberg, J. R. (2009). Changes in dynamic trunk/head stability and functional reach after hippotherapy. *Archives of Physical Medicine and Rehabilitation*, 90(7), 1185–1195. https://doi.org/10.1016/j.apmr.2009.01.026

Silkwood-Sherer, D. J., Killian, C. B., Long, T. M., & Martin, K. S. (2012). Hippotherapy-an intervention to habilitate balance deficits in children with movement disorders: a clinical trial. *Physical Therapy*, 92(5), 707–717. https://doi.org/10.2522/ptj.20110081

Silkwood-Sherer, D., & Warmbier, H. (2007). Effects of hippotherapy on postural stability, in persons with multiple sclerosis: a pilot study. *Journal of Neurologic Physical Therapy: JNPT*, 31(2), 77–84. https://doi.org/10.1097/NPT.0b013e31806769f7

Stergiou, A., Ploumis, A., Beris, A., Ntzani, E., Tzoufi, M., & Varvarousis, D. (2017). Therapeutic effects of horseback riding interventions: a systematic review and meta-analysis. *American Journal of Physical Medicine & Rehabilitation*, *96*(10), 717–725. https://doi.org/10.1097/phm.0000000000000726

Tseng, S., Chen, H., & Tam, K. (2012). Systematic review and meta-analysis of the effect of equine assisted activities and therapies on gross motor outcome in children with cerebral palsy. *Disability and Rehabilitation*, 35(2), 89–99. https://doi.org/10.3109/09638288.2012.687033

Zadnikar, M., & Kastrin, A. (2011). Effects of hippotherapy and therapeutic horseback riding on postural control or balance in children with cerebral palsy: a meta-analysis. *Developmental Medicine & Child Neurology*, 53(8), 684–691. https://doi.org/10.1111/j.1469-8749.2011.03951.x

Žalienė, L., Mockevičienė, D., Kreivinienė, B., Razbadauskas, A., Kleiva, Ž., & Kirkutis, A. (2018). Short-term and long-term effects of riding for children with cerebral palsy gross motor functions. *BioMed Research International*, 2018, 1–6. https://doi.org/10.1155/2018/4190249

Made in the USA
Monee, IL
08 November 2026

68074903R00057